MW01248517

FINDING PLACES

The Search for the Brain's GPS

FINDING PLACES

The Search for the Brain's GPS

Unni Eikeseth

Norwegian University of Science and Technology, Norway

Translated by **Lucy Moffatt**

NEW JERSEY • LONDON • SINGAPORE • BEIJING • SHANGHAI • HONG KONG • TAIPEI • CHENNAI • TOKYO

Published by

World Scientific Publishing Co. Pte. Ltd.
5 Toh Tuck Link, Singapore 596224
USA office: 27 Warren Street, Suite 401-402, Hackensack, NJ 07601
UK office: 57 Shelton Street, Covent Garden, London WC2H 9HE

British Library Cataloguing-in-Publication Data
A catalogue record for this book is available from the British Library.

This translation has been published with the financial support of NORLA, Norwegian Literature Abroad

NORLA
NORWEGIAN LITERATURE ABROAD

Unni Eikeseth: *Jakten på stedsansen. Hvordan May-Britt og Edvard Moser løste en av vitenskapens store gåter*

Sold through Winje Agency A/S, Skiensgate 12, 3912 Porsgrunn, Norway

FINDING PLACES
The Search for the Brain's GPS

ISBN 978-981-121-691-6 (hardcover)
ISBN 978-981-121-692-3 (ebook for institutions)
ISBN 978-981-121-693-0 (ebook for individuals)

For any available supplementary material, please visit
https://www.worldscientific.com/worldscibooks/10.1142/11735#t=suppl

Typeset by Stallion Press
Email: enquiries@stallionpress.com

Printed in Singapore by Mainland Press Ptd Ltd.

Contents

CHAPTER 1

An Unerring Instinct

White, white, white. The eyes of the Russian lieutenant Ferdinand von Wrangel and his men stung and ached in the sharp light reflected off the snow-covered sea ice. It was early April 1821. Just over a week earlier, the expedition had left the mainland and set off north across the frozen sea in 21 sledges drawn by a total of 240 dogs. For the first few days, they could still see the Baronov cliffs of the mainland on the horizon behind them. After that, the world around them became one endless icy plain, relieved only by open channels and pack ice.

Von Wrangel was leading one of the two Siberian expeditions dispatched by Czar Alexander. His mission was to map the coast of Siberia in exhaustive detail. The party was also supposed to investigate whether there was any uncharted territory in the Arctic Ocean north of Siberia, as other explorers claimed.

For von Wrangel and his men, being able to identify where they were and find their way back across the frozen sea, without landmarks by which to navigate, was a matter of life and death.

When open channels and large blocks of ice forced the expedition to switch direction, the lieutenant would set a course. Von Wrangel was aided by the most modern navigational instruments then available, including two chronometers, which served as portable time standards. In addition, he had a stopwatch, a sextant and an artificial horizon, three azimuth compasses, two telescopes, and a measuring line. Toward the end of each day's leg, he would collate all the readings to determine the expedition's precise position. Remarkably, several of the experienced local sledge-drivers were far better than he was at calculating their position after a day's travel across the ice, even without recourse to any navigational instruments. "They appeared to be guided by a kind of unerring instinct," the lieutenant wrote in his account.

Von Wrangel was particularly impressed by a seasoned sledge-driver and guide, Sotnik Tatarinow:

> In the midst of the intricate labyrinths of ice, turning sometimes to the right and sometimes to the left, now winding round a large hummock, now crossing over a smaller one, among such incessant changes of direction, he seemed to have a plan of them all in his memory and to make them compensate each other, so that we never lost our main direction, and whilst I was watching the different turns, compass in hand, trying to resume the route, he had always a perfect knowledge of it empirically. His estimation of the distances we had passed over reduced to a straight line, generally agreed with my determinations deduced from observed latitudes and the day's course.[1]

After nearly getting trapped in rotten, melting sea ice, von Wrangel's expedition made its way back to the mainland by the skin of its teeth. They hadn't found any land to the north, but they had discovered several islands and filled in a few blanks on the map along the Siberian coast. Several decades after von Wrangel's arduous journey, his account was published in English and read by the renowned evolutionary biologist, Charles Darwin. Darwin noted the description of the sledge-drivers' incredible sense of direction and began to brood over how this could be possible.

In his twenties, Darwin had himself been on a long expedition aboard *The Beagle*. He was well aware how challenging it was to keep track of one's journey in the conditions that von Wrangel and his men had navigated. If you were constantly obliged to change course, neither a compass nor the Northern Star would be sufficient for determining your position on the open sea. The sledge-drivers must have been tapping into some subconscious ability to calculate speed, distance, and time. Yet, Darwin did not believe that the sledge-drivers had a special faculty. He suspected that all humans had an overview of where they were; the sledge-drivers simply appeared to have perfected this ability. Darwin figured this sense was influenced by sight, and perhaps information from one's own muscle movements.

And so Darwin formulated an idea that remained unproven long after his death. Could it be that some part of the brain was specialized for determining direction?[2]

It would be another 130 years until this question was answered by a husband-and-wife research team at a small university in Norway.

CHAPTER 2

Hareid Planet Club

One afternoon in fall 1982, a chance meeting took place on an Oslo street. Thirty-two years later, two of the people who met would dominate the news for weeks on end when they received Norway's first ever Nobel Prize in Physiology or Medicine.

Nineteen-year-old May-Britt Andreassen had just finished her shift at the Kaffistova café and was on her way down toward Karl Johans gate when she spotted a couple of familiar faces from her hometown. The people in question were Øyvind Strand, who had attended the same high school as her in Ulsteinvik, and Edvard Moser, from her chemistry class. Edvard told her he would be studying science in Oslo in the spring. May-Britt, who had been studying in the city for half a year, immediately offered to show Edvard around. She remembered what it was like when she was a newly fledged student, how strange and new everything was, and was keen to help him get to know the university and the city.

May-Britt didn't know Edvard particularly well, and her impression from their schooldays was that he seemed shy. But a few months later, he returned to Oslo and surprised her by getting in touch. He was eager to be shown around Blindern campus.

The two had grown up about 20 kilometers apart in Sunnmøre, northwestern Norway, she on a smallholding in Fosnavåg on Bergsøya, and he in the village of Hareid, on the neighboring island of Hareidlandet. May-Britt was the baby of her family while Edvard was the eldest brother in a family distinguished by an unusual story. His parents were German and had come to Norway in the 1950s when his father got a job at an organ factory on the island of Haramsøya.

They both started school in August 1969, just a few weeks after Neil Armstrong and Buzz Aldrin became the first people to land on the moon in Apollo 11. Although for seven-year-olds from Sunnmøre the United States and the moon were both remote from their everyday lives, all the first-graders knew it was possible for humans to make it so far they could leave a footprint on an alien planet. Like many children in the era of moon landings, Edvard Moser was interested in astronomy and space travel. Together with his schoolmate Øystein Orten, he had founded the Hareid Planet Club, whose members exchanged facts about the solar system and the distances between the planets. But he had many other interests as well, among them collecting stones and setting up chemical reactions in the bathroom.

May-Britt, an energetic child, had tried out most of the activities on offer in her neighborhood. Like many other children in Norway's Bible Belt, she went to Sunday school. Later, she took up swimming and mountain hiking, joined the Scouts, and played guitar. She could often be found in the company of her dog Bamse, a big Norwegian buhund her parents had given her.

During Edvard's early months in Oslo, the two became good friends, bonding over the fact that neither could decide what to study. Edvard had tried inorganic chemistry but quickly discovered that chemistry wasn't for him. He considered studying other hard sciences, but a book he was reading at the time gave him a taste for something completely different. He had become fascinated by Sigmund Freud's *The Interpretation of Dreams*, in which Freud described both his own dreams and those of others, claiming that they were a gateway to the innermost psyche of human beings.

"I wanted to understand why people acted the way they did, and why they dreamed the way they did, and whether Freud was right to believe dreams offered us insight into the workings of the human mind. That's what sparked my interest in the subject of psychology," Edvard Moser says.[3]

Although May-Britt shook her head at the idea of Edvard reading something so unscientific, she shared his enthusiasm for psychology. Having lived in close contact with nature and animals since childhood, she too had developed an interest in why animals and humans behaved the way they did.

That is how, in autumn 1983, the pair came to be sitting together in the lecture hall of Oslo's psychology department. They discovered early on that one particular branch of psychology appealed to them more than the rest: biological psychology, a discipline in which researchers used scientific methodology to reveal the underlying biology of human and animal behavior. In a sense, something clicked for them both: this was exactly what they wanted to study.

Their lecturer Carl Erik Grenness gave them a special edition of a book titled *The Brain*, which had been published by *Scientific American*. The book introduced them to recent breakthroughs in biological psychology and neurobiology.[4]

In the decades before Edvard and May-Britt began their studies, psychology had undergone a revolution, propelled by advances in brain science, biology, and chemistry. The pair realized that they were students in an era when psychologists were just beginning to reveal how the brain learned, and how that was reflected by animal and human behavior. It was known that learning involved contact between different neurons, which "talked" to each other at their points of contact, the *synapses*. Memory entailed the re-activation of the same neurons that had been implicated in learning.

In *The Brain*, they read, among other things, about the groundbreaking experiments the American scientist Eric Kandel had carried out on the nervous system of the *Aplysia* slug. Despite its primitive nervous system, the slug could still learn. Its neurons are among the largest in the animal world – one millimeter in diameter. They are visible to the naked eye, which makes the slug a particularly suitable research subject.[5]

Aplysia has a natural reflex that prompts it to withdraw the gills on its underbelly when threatened by danger. Using this reflex as the starting-point for his experiments, Kandel trained *Aplysia* to distinguish between harmless touch (*habituation*) and touch to which it was supposed to be particularly sensitive (*sensitization*). When the slug was exposed to the latter, it simultaneously received an electric shock. Kandel discovered that the two different types of learning – habituation and sensitization – manifested themselves differently in the neurons. Neurons that had become more sensitive grew more synapses onto the cells that controlled the gill reflex than those that had become used to the touch being harmless. This was the first concrete example of learning leaving physical traces on the nervous system.[6]

Edvard and May-Britt were disappointed when they discovered that there was no community of biological psychologists in Norway. But they soon discovered that Norway had something almost as interesting: The University of Oslo had a world-renowned community of neuroscientists who were working on one of the most exciting areas of the brain when it came to memory and learning: the *hippocampus*.

CHAPTER 3

Hippocampus

The first scientists to study the brain noticed the hippocampal structure, located several centimeters beneath the temple, in the temporal lobe (Figure 1). The region is striking in form – reminiscent of a seahorse – and so in 1587 the Italian anatomist Giulio Aranzi named it after the Latin word for seahorse, *hippocampus*. Before then, the structure had been known as the *Cornu ammonis*, after the ancient Egyptian God Amun, with his curved horns. Another aspect that bolstered interest in the structure was that it could be found in the brains of all mammals, from the most primitive to the most highly developed. The hippocampus, therefore, had to have considerable and fundamental significance.

When the earliest scientists started to experiment with dyeing thin slices of brain at the end of the 1800s, they discovered that the offshoots in the hippocampus were incredibly neat and tidy. First came one region of a cell type with branches that reached out to a new region of cells and then a third region after that. The three different regions of cells were named dentate gyrus, CA1 and CA3. The abbreviation CA came from the old name for the hippocampus, *Cornu ammonis*. The dyed neurons revealed that the hippocampus was neatly organized and, at the same time, that it had considerable connections to other areas of the brain. The structure was evidently extremely important – but in what sense?

In 1953, scientists got a hint of what it might be. In August of that year, a young man lay on an operating table in the US. His name was Henry Gustav Molaison, and later he would become known to the world's neuroscientists as H.M. From the age of 10, Molaison had suffered epileptic seizures. The disease may have been caused by a head injury he incurred during a cycling accident at the age of nine, which left him unconscious for five minutes. But it may also have been hereditary. Several family members on his father's side suffered from epilepsy.

From the age of 16, Henry Molaison started to have more severe seizures. He would faint, wet himself, bite his tongue, and suffer terrible convulsions of the arms and legs. Despite all this, Molaison made it through high school and went on to work as an auto mechanic and then at a typewriter factory. To ease the seizures, he took large doses of anti-epilepsy medicine. But soon the disease forced him to stop working, and by the time he was 27, the situation was so desperate that he and his family were willing to try surgery.

Many epilepsy patients had showed improvement after small parts of the brain where the seizures originated were removed. The problem in Henry Molaison's case was that the doctors hadn't managed to identify a precise target for the surgery, even after taking a series of measurements from electrodes attached to his head. The surgeon William Beecher Scoville therefore suggested an experimental operation that had previously been carried out only on patients with psychiatric disorders such as schizophrenia. The surgery involved removing parts of the medial temporal lobe in both halves of the brain. This is where the seizures originate for a large proportion of epilepsy patients, so there was hope that Molaison too might see improvement after such an operation.

Molaison would then lack both sides of his hippocampus, but there was nothing to suggest that he wouldn't be able to get by without them. At the time, little was known about the function of the hippocampus. A British scientist summed up the knowledge of the times as follows: "The striking aspect of the hippocampus is the anatomical elegance of its structure, revealed in detail in the past few years. In contrast, there is really appalling ignorance about what this elegance means."[7]

Yet there were some hints. In 1888, two physiologists at the University of London, Sanger Brown and Edward Schäfer, tried to identify the functions of the different parts of the temporal lobe by experimenting on rhesus monkeys. In one of the monkeys, described as large and active, they removed the entire temporal lobe and the hippocampus. After the operation, all of the monkey's senses appeared to be intact, but the scientists noticed a peculiar change. The monkey scrutinized objects, other monkeys, and researchers each time it saw them, even when it had last seen them only a few minutes before. It was "as if it had completely forgotten its earlier experiences."[8] In 1900, a Russian neurologist, Vladimir Bechterev, also described a patient who had severe memory problems. When the patient died, the autopsy showed damage to the brain tissue in the temporal lobes and hippocampus.[9] These examples were, however, deeply buried in research literature, and few psychologists and brain surgeons knew about them or understood their significance.

On Tuesday, August 25, 1953, Henry Molaison lay awake on the operating table, talking to Scoville and the nurses around him.[10] He had been given only local anesthetic in his forehead, where the surgeon would make the incisions. The brain itself has no pain receptors, so there was no need for a general anesthetic. As soon as the local anesthetic had taken effect, Scoville made an incision along a wrinkle in Molaison's forehead and drew the skin aside, revealing the bone beneath. Then, Scoville bored two holes in his skull, just above the eyebrows, and removed the two round fragments of bone from the holes, bringing the brain membrane into view.

Scoville made one last attempt to identify the origin of Molaison's seizures. He placed electrodes directly into the exposed brain tissue, but this time, too, he failed to find the seat of the seizures. So Scoville continued with the surgery. He cut through the tough brain membrane, revealing the brain itself. Molaison's brain pulsated in time with his heartbeats and breathing. Scoville inserted a long brain spatula in through one of the holes and lifted up the frontal lobe. This simultaneously caused the rest of the brain to sink a little farther down into the skull, giving him more room for maneuver. He could now see the hippocampus. The surgeon inserted a new instrument into the hole in the skull to suck out bits of the soft brain mass. Bit by bit he sucked out around half the hippocampus and some of the nearby cerebral cortex. After that, he did the same thing on the other side of the brain.

Shortly after the operation, it became clear that the surgery had had an unexpected and tragic effect on the young patient. Henry Molaison no longer recognized the hospital staff, was unable to find his way to the toilet, and didn't seem to remember anything that had happened during his hospital stay. Only his childhood memories appeared to be intact. It was as if he was experiencing everything for the first time.

This dramatic effect convinced many scientists that the hippocampus was significant for human memory.[11] Nonetheless, it was not seen as conclusive evidence. One important reason for this was that a fairly substantial part of Molaison's temporal lobe, including the amygdala and the tissue around the hippocampus, had also been removed. It was possible that the absence of these other areas might also have contributed to Molaison's inability to create new memories.

In subsequent years, interest in hippocampus research surged worldwide. Since, of course, human brain research was prohibited, scientists had to find ways of studying memory in animals. But how can you tell whether an animal has learned something when it can't tell you that itself? Researchers discovered that rats' natural capacity for finding their way through mazes was an ideal starting-point for investigating the role the hippocampus played in learning. Although the rats couldn't tell the researchers what they had learned, it was easy to measure how long the rats took to find their way to a particular place in a maze.

Still, psychologists were a long way from understanding the processes that occurred in the human brain when new memories were created, or when the brain was learning something new. Everyone hoped to catch the mammal brain in the act – to see a physical trace of learning in its neurons.

CHAPTER 4

The Trace of Memory

By the early 1980s, when Edvard and May-Britt were students, brain research had become so popular that the Norwegian Broadcasting Corporation (NRK) created a TV program about it, *Your Fantastic Brain*, and broadcasted it at prime time. This was largely thanks to Per Andersen, an internationally renowned neuroscientist and a gifted communicator.

Each program started with a darkened image of a brain and the silhouettes of two men at a table. Then the lights went up, and the camera panned into the presenter, Per Øyvind Heradstveit, wearing a suit and glasses with square black frames. Heradstveit launched one episode on memory like this: "Today we'll be dealing with memory recall, in other words, the ability to remember. But before we can remember something, we must, of course, have thought something or learned something. So what actually happens when we think?"

The TV image switched to Per Andersen, also wearing a suit and tie, and apparently even more at home in the studio than the presenter himself. He smiled, revealing a small gap between his front teeth: "Well, that's quite the question. I could ask you the same thing." Andersen explained that scientists still didn't know exactly what happened, but that people believed a thought was a kind of relay race during which many neurons sent impulses to one another. If you switched out some of the neurons in the relay chain, you would get a different thought.[12] The TV series gave Norwegians insight into the very latest research on the brain and transformed Per Andersen into a national celebrity.

Now Edvard and May-Britt suddenly found themselves sitting, almost star-struck, in the same lunchroom as this TV personality. They didn't dare speak but just listened to the exciting stories of Andersen and other senior faculty in the department.

At this point, in 1988, Edvard and May-Britt were approaching the end of their psychology studies and had long since gone from being good friends to a couple. They did everything together. They had been on vacation together in South America and Africa. In 1984 they had gotten engaged at the top of Africa's highest peak, Kilimanjaro, and a year later they were married. As part of their psychology studies, they were both working on a project headed by Terje Sagvolden, a psychologist and colleague of Per Anderson's in the neurophysiology department. Sagvolden's laboratory was based in one of the venerable old university buildings across from the National Theater on Karl Johan, Oslo's main street.

The project involved studying how a particularly hyperactive type of rat differed from non-hyperactive rats. The idea was that if you understood why hyperactivity occurred in rats, you could, perhaps, understand ADHD in humans. Together with Sagvolden, they had published three articles about behavior in hyperactive rats, which they were proud of. Even so, they had a nagging feeling that they weren't in the right place. What they really wanted was to study the changes that occurred in the brain when an animal learned, and they wouldn't get to do that with Sagvolden. And so they began looking for a new research community.

A seminar that was held at the department proved decisive, giving them insight into an important breakthrough that had been made in Andersen's research group – a learning mechanism that became known as long-term potentiation.

Per Andersen was an expert on the hippocampus. His research group was working to map what was seen as the basis of learning among animals and humans: how the neurons in the hippocampus communicated with one another through electrical signals and how the connections between neurons were strengthened.

Andersen and his research group were known for several groundbreaking discoveries. In the 1960s, while Andersen was working with the Nobel Prize winner John Eccles in Australia, he discovered the direction in which neural signals travel through the hippocampus.

Until then, people had believed that the signals traveled from an area in the upper section of the hippocampus, CA1, to another region, CA3, at the bottom of the structure. This was what the father of neuroscience, the physician and artist Santiago Ramón y Cajal, had sketched into his otherwise flawless anatomical drawings of the brain. But based on his own studies of the hippocampus,

Andersen suspected Cajal might have been mistaken. He thought that the signals went in the opposite direction.[13]

Consequently, Andersen designed an experiment intended to reveal which hypothesis was correct. Working with a couple of colleagues, he inserted an electrode into the brain of an anesthetized rabbit and stimulated the main input to the hippocampus, the so-called perforant pathway, with a mild electrical current. In order to find out how the signal spread from cell to cell in the hippocampus, they also set up measuring points in the three regions of cells in the hippocampus: the dentate gyrus, CA3, and CA1. Upon activating the perforant pathway, they found that the cells in the dentate gyrus were activated, followed by cells in CA3, and then cells in CA1, exactly as Andersen had predicted. In order to be absolutely certain, they also cut the fibers between the CA3 and CA1 and stimulated the perforant pathway. This time, they saw that the activity in CA3 was unchanged, while cells in the CA1 region remained silent.

This was the conclusive proof that the signal traffic in the hippocampus traveled in the direction Andersen had thought. Because the signals passed over three regions, it was called the trisynaptic circuit (Figure 2).

Just after Per Andersen returned to Oslo, his research group made another major discovery. His PhD student, Terje Lømo, had stimulated neurons in the hippocampus with electrical signals and found that these neurons reacted more rapidly and far more strongly when they were exposed to similar impulses at a later date.[14] Remarkably, the amplified effect could last for many hours – something nobody had seen before. It looked as if the electrical impulses caused a lasting change in the neurons. The cells could "remember" what they had experienced before! Lømo and Andersen called the effect long-lasting potentiation. It was later renamed long-term potentiation, or LTP.

The effect is similar to the way our own reactions are sometimes amplified by an experience. Imagine that your sister always makes sarcastic comments. You are irritated but mostly manage to ignore them, until one day she crosses the line with an especially nasty comment, which makes you explode and then retaliate. Before you have time to calm down again, she comes out with a fairly harmless gibe, but because of what has just happened, you start firing on all cylinders. The comment that crossed the line makes you react much more strongly to the innocent remark that you normally would have ignored. Your response has been amplified, or potentiated.

Lømo and Andersen quickly realized that this amplification could be the kind of plasticity, or malleability, that people had envisaged as the basis for learning in the brain. Even so, for the first 10 years after the discovery, only a few research groups followed up on this finding.[15] But in the eighties, interest in the LTP phenomenon surged again, bringing about a series of important findings that suggested LTP could be a learning mechanism in animals.

In 1986, Richard Morris and his colleagues at the University of Edinburgh discovered that when they blocked LTP-like processes in rats' brains using chemical substances known as antagonists, the rats were no longer able to learn their way to a platform in a water tank – an activity rats usually performed with ease.[16] This suggested that LTP was, in fact, reminiscent of the natural learning process in animals.

"The fact that it's possible to see changes in the brain as a result of learning, that was just . . . we were so stoked," says May-Britt Moser.[17]

The question was whether this long-term potentiation actually played a role in learning and memory. Edvard and May-Britt wanted an answer. And so they decided to contact Per Andersen and ask to be accepted into his research group.

CHAPTER 5

Swimming Rats

One day in 1988, Edvard and May-Britt Moser knocked on the door to Per Andersen's office.

"He was a person people respected a great deal, including us. So it took a lot of nerve to work up the courage to go to his office. But we were desperate. Our future totally depended on whether or not he agreed to supervise our master's thesis," says May-Britt Moser.

As they had feared, he started off by informing them he didn't have room for any more students in his research group. But May-Britt was determined not to leave until he agreed.

A damned nuisance, Per Andersen thought. He didn't really mean it in a negative way. In his three decades leading a research group, he had learned that the most talented students were often the most determined. They had drive – they were after something. The Mosers clearly fell into this category. Andersen had heard about the Mosers and knew that they were skilled and enthusiastic workers. Since it was clear that they wouldn't give up easily, he decided to give them a chance. At the same time, he wanted to test their mettle.[18] So he took out an article by Richard Morris, which described the water maze Morris had developed to test spatial memory in rats. Andersen said that if they wanted to work on their master's thesis under him, they would have to read this article, understand it, and replicate the water maze experiment.

"Fantastic, because we want to do our PhDs under you as well!" May-Britt exclaimed.

When they came out of the meeting, they were happy but also somewhat intimidated by the task they'd been assigned. They needed a water tank that was two meters in diameter and half a meter high, large enough that a Plexiglass platform could be submerged in the water. Andersen, a keen sailor, thought they should build it from scratch using the same materials that were used to build sailboats. But after a telephone call to May-Britt's brother-in-law, who worked in the shipping certification body, Veritas, they discovered that a firm in Sunnmøre produced tanks of that size, which they could buy readymade.

They set up their own laboratory in a little room in the basement of the neurophysiology department. Since they had not yet finished their psychology degrees, they spent their days at lectures, then worked on their laboratory into the night. The tank they had obtained held 330 gallons of water, and they filled it by hooking a long hose up to the faucet in the bathroom across the corridor. On Andersen's advice, they bought a boat pump to empty the water into the bathroom.

They installed a camera in the ceiling, positioning it over the center of the pool so that they could observe the rats. Bruce Piercey, a computer programmer who worked in the department, designed software to track the rats' movements. Since it was crucial that the rats not be able to see the bottom of the pool or the transparent Plexiglass platform, they added milk to the water to make it opaque. To prevent the milk turning sour and starting to smell bad, they had to change the water every day. The task of filling the tank in the morning and emptying it in the evening took several hours. But that didn't matter to them: The most important thing was that they had gotten their own lab.

At first they used white albino rats in their experiments, but the white bodies blended in with the milky white water. For a while they solved this problem by painting the rats' heads with water-resistant mascara, before eventually getting hold of hooded rats, which had black heads and were much easier to track. The hooded rats also had better vision than the albino rats, and found it easer to orient themselves in the pool.

The Mosers wanted to determine whether it was possible to observe learning in rats' hippocampi. However, there were too many cells in the hippocampus to have any hope of finding traces of learning; it would be like searching for the proverbial needle in the haystack. Andersen suggested removing all but a small part of the hippocampus so that they would have a smaller area to search. Edvard and May-Britt agreed on this but realized that they first needed to check whether it was even possible for rats to learn without a fully intact hippocampus. After all, nobody knew whether the entire hippocampus contributed to spatial memory in the same way or whether some parts were more involved than others.

Unsurprisingly, answering this question became the topic of their master's thesis.

Andersen taught them how to perform brain surgery on the rats. They were also helped by Per Andersen's collaborator of many years, Theodor Blackstad – among the world's foremost experts in the anatomy of the hippocampus. Blackstad showed them which "landmarks" to look out for before entering the brain and where the boundaries lay between the hippocampus and the surrounding tissue.

Since Per Andersen had many international collaborators, world-famous researchers such as the Nobel prizewinners John Eccles and Eric Kandel often visited Oslo. These guests were always honored with festive gatherings either at Andersen's holiday cabin in Hemsedal or at the Andersen home at Blommenholm, Bærum, on the outskirts of Oslo. There, Andersen and his wife Kari Sletten prepared marvelous dinners for as many guests as they could fit around the table. As a rule, Andersen drew the line at PhD students: if master's students were also included, there simply wouldn't be room for everybody. But now and then he made an exception, inviting May-Britt and Edvard. As a result, they got to meet some of the world's foremost neuroscientists.

Their experiments started in January 1989. First they anesthetized a rat and placed it on an operating table, where it would receive constant top-ups of anesthetic gas throughout the operation. After that, they carefully drilled holes in the skull and sucked out parts of the hippocampus on either side of the rat's brain. In some rats, they removed the lower part of the hippocampus, and in others, the upper part. They also had control groups of rats which hadn't had any of their hippocampus removed but either simply had the same small holes drilled in their skull as the other rats or were missing part of their cerebral cortex. The control animals were important for ensuring that the researchers were recording the effect of the missing area of the hippocampus, and not some other factor.

After the operation, the rats were left to rest in their cages for a week before being trained in the pool. There, they were supposed to find the Plexiglas platform that lay hidden a few centimeters below the surface, concealed by the milky white water.

One by one, the rats were carefully lowered into the tepid water, their heads facing the edge of the pool. As soon as each rat was released, it started swimming around and searching for a foothold – natural behavior in rats. Upon locating the platform, each rat was left to stand there for half a minute before being removed from the tank. The rats that hadn't managed to find the platform after two minutes were helped to find their way. After a short break, the rats were lowered into the tepid water a second time, but this time in a different area of the pool. After that they were left to rest for four hours before it was time for a new session.

This went on for eight days.

During this time, the rats in the control group learned how to find the platform in fewer than 12 seconds. The rats that were missing the lower part of the hippocampus also performed well in the test. But things were different for the rats that were missing the upper area of their hippocampus. They took much longer to find their way to the platform. This was a surprising finding. It indicated that the upper part of the hippocampus played a more important role than the lower area in enabling rats to orient themselves. But the Mosers couldn't explain why this was the case.

Several months after handing in their master's theses, the Mosers presented their research at a European neuroscience conference in Stockholm. One of the people who stopped by their poster was Richard Morris, the scientist who had invented the water maze they used in their experiments. Morris praised them for their interesting study. He was amazed that they had struck on the idea of removing tissue from different parts of the hippocampus; he had never thought of that. But he also pointed out that there was a risk of damaging neurons in other parts of the hippocampus if they had connections that passed through the area that had been removed. Morris invited them to Edinburgh, where they could learn to use a less intrusive method that wouldn't damage these neural connections. They were thrilled. One of the truly great scientists in the field was interested in what they had done and had invited them to work with him!

Now they were ready to continue their search for traces of learning in rat brains.

CHAPTER 6

A Warm Brain

The committee assessing PhD applications in psychology had a tough job in 1990. Only five grants were on offer in the whole of Norway. The tricky thing was that two of the most qualified applicants were from the same research community in Oslo, shared the same supervisor, and, on top of it all, were married to each other. It would be controversial to award almost half that year's resources to the very same research group, but these two applicants were exceptional. Two of the committee members were especially enthusiastic and worked hard to convince the others: Kenneth Hugdahl, a professor at the University of Bergen, and Svein Magnussen of the University of Oslo.[19] After a long discussion, the committee ultimately decided to award both of the married researchers a grant.

Although it had been 20 years since long-term potentiation (LTP) was first discovered in Per Andersen's lab, nobody had found conclusive proof that this kind of persistent strengthening of neural signals occurred when animals and humans learned in real life. To prove that LTP did in fact underlie learning, they would have to measure LTP in freely moving animals, and this was a challenge Edvard was ready to take up. To do so, he would have to surgically implant permanent electrodes into rats' brains. Up until then, he and May-Britt had simply surgically removed parts of the rats' hippocampi and observed how that affected their ability to learn; but they had no experience surgically implanting electrodes and recording impulses from the neurons of living animals. Since nobody else in Andersen's laboratory knew how to do this either, they turned to Boleslaw "Bolek" Srebro for assistance.

Srebro had worked on the hippocampus at a prestigious neuroscience research institute in Poland, the Nencki Institute. In 1969, he had moved to Norway to start work with a newly launched neuroscience group at the University of Bergen.

Edvard and May-Britt were invited to Srebro's lab, where, over the course of a week, he taught them how to make the tiny electrodes that they would surgically implant in the rats' hippocampi, how to perform the operation, and how to read the neural impulse recordings in awake rats.

Neurons differ from other cells in that they "fire" electrical impulses. At rest, a neuron is negatively charged in relation to the fluid outside the cell. But when the cell reacts to something, this situation is turned on its head. In a fraction of a second, the inside switches from being negatively charged to positively charged. This is called an action potential, a nerve impulse, or a spike. The impulse can be measured by placing thin electrodes into the tissue immediately outside the cell, or directly into the cell – although the latter is a lot more difficult. Since the change in polarity is so brief, 100 times quicker than the blink of an eye, scientists use a special instrument to measure the electric signal: an oscilloscope. In an oscilloscope, a stream of electrons flows in a straight line across a screen, with changes in polarity visible as crests or troughs above or below the line.

Edvard planned to record nerve impulses in the tissue outside the neurons, which meant he would obtain a composite electrical impulse from many brain cells. These kinds of impulses are called a field potential. If the field potential was enhanced, this would indicate that the synapses between the neurons had been potentiated and that the cells had learned. It is only possible to measure these sorts of field potentials if the neural pathways are first stimulated with electricity. So they had to implant both an electrode that would supply current and electrodes that would detect the neural responses.

Upon returning to Oslo, Edvard started to measure field potentials in the hippocampus of rats. He started by recording on "dry land," in an open box on the floor. At first, the results looked promising: as expected, when a rat was exploring its surroundings, the electrical signals were enhanced compared to when it was at rest. This had been observed by others. But when Edvard carried out the same recordings on animals in the water maze, something unexpected happened. Instead of the spikes becoming enhanced, they decreased. This was unexpected because the rats were obviously learning the task.

Meanwhile, May-Britt took a bit more time to find a PhD project. Since they had agreed that Edvard would work on LTP, she had to find another project that would demonstrate how learning occurred. Andersen made a suggestion. He had a colleague in the toxicology department who was interested in researching whether alcohol destroyed the synapses between the neurons in animals' hippocampi. May-Britt wasn't fired up by this project. For one thing, she was extremely averse to the idea of giving the rats the necessary amounts of alcohol. For another, she thought the resultant damage would be too non-specific to give them any meaningful data. She managed to get out of it by telling Andersen that because she was from the Bible Belt she simply didn't want to give alcohol to rats.

She came up with another idea. Instead of seeing what might destroy synapses, she wanted to try to study the opposite. She would train animals in an enriched environment and observe whether they developed more synapses between their neurons than rats that grew up in a less stimulating environment. Andersen had just invested in an expensive, state-of-the-art confocal microscope, and she thought she could use this microscope to see whether the neurons developed more synapses as the animals learned. She wrote up a project outline and presented it to Andersen. He didn't believe it would be possible to observe such changes using the microscope and initially advised her to forget the idea. But May-Britt stood her ground, and eventually he agreed.

Per Andersen's research group had long attracted the very brightest students, but it was perhaps at its peak in this period. Edvard and May-Britt were surrounded by a group of highly motivated, hard-working, and self-driven scientists. Apart from them, the group consisted of four physicians: Mari Trommald, Morten Raastad, Ole Paulsen, and Paola Pedarzani. All of them practically lived in the department, which had by now moved to a new building in Gaustad. As a rule, they went home only to sleep. Cooking wasn't high on their list of priorities, although now and then Pedarzani made authentic Italian pizza at work.

Even though all the PhD scholars were working on different projects, they regularly discussed their problems with one another. They also had a regular weekly meeting, a "journal club," during which students took turns presenting an article that they had read. Afterwards, they would all discuss the contents. Since they had different academic backgrounds, they could learn a lot from one another. Nobody prepared more thoroughly for the article reviews than Edvard Moser, who would read the articles several times, armed with a ruler and different-colored markers. The first time around he marked important parts in yellow. Then he read the article again, highlighting parts of the text in green. Finally, he used a red marker. The others noticed that he always identified the particularly important aspects of a study.

The results of his water maze research continued to frustrate Edvard. He started to test different variables to see if something other than learning could be influencing the recordings. When he started to look over his results, he realized that the field potentials were most enhanced when the rats were swimming in warm water and decreased considerably in cold water. Could water temperature be a contributing factor? Was the phenomenon that he and so many other scientists had taken for proof of learning a temperature effect? To be certain, he had to test this systematically.

Yet, research was not the most important thing in Edvard and May-Britt's life in that period. Six months after they started their PhDs, May-Britt became pregnant. On June 8, 1991, she and Edvard became parents to a daughter, Isabel.

Isabel often came to work with her parents. She was a contented child who was happy to sit in a playpen or crawl around on the floor exploring as her parents worked.

"Edvard and I were so naïve that we didn't even ask for permission to bring a child to the office. It was so obvious to us that it must be allowed because she wasn't doing anything wrong and she was so good and never cried. Per probably heard about it from the other professors but he shielded us from the worst of the criticism. In a way, he must have seen and accepted our need to work hard, even though he himself was a traditional family man," May-Britt Moser says.

May-Britt considered her PhD work to be very child-friendly. A lot of her time was spent making colorful toys for the rats or altering their cages by adding new levels and the like. Isabel often sat on her lap and watched the funny rats as they sniffed about and explored their cages.

While Isabel got steadily better at exploring her world, learned to crawl and stand up, and then took her first step, Edvard and May-Britt worked to understand how the cells in the hippocampus supported learning. Edvard had been forced to make a major detour: he was still trying to grasp how water temperature affected his recordings. Andersen suggested inserting a tiny thermometer, a so-called thermistor, in the rats' brains.

They asked the technician Eva Aaboen Hanssen for help, and she set to work making hair-thin temperature gauges. Iacob Mathisen, a master's student, was also on hand to help with the recordings.

Until then, it had been thought that the brain had mechanisms for maintaining its temperature at a constant level. However, the results of the thermistor readings showed that brain temperature was highly dependent on external temperature. They also found that there was a direct link between brain temperature and the field potentials. In water maintained at 64.4 degrees Fahrenheit, the brain temperature rapidly dropped around 10 degrees, and the field potentials simultaneously decreased. If the rat was moving or placed under a heat lamp, however, the brain rapidly heated up and the field potentials were enhanced. The temperature effect was substantial and could mask potential learning-related effects.[20] The finding was published in *Science*, one of the world's most reputable scientific magazines, and came as an unpleasant shock to many scientists, who had viewed an increase in field potentials as a sign of learning.

"That was a surprise, and that got our attention," says Carol Barnes, one of the scientists who had done a lot of work on field potentials in animals.

"We were coming around to that idea ourselves, but we didn't get the thermal probes in as the Mosers did. Your brain temperature can vary by a couple of degrees just from walking or running around, so the dogma in the field is that the hypothalamus keeps the brain temperature very stable. It doesn't."

"Morris's water maze was used in a huge number of experiments to study aging, brain damage, the effects of drugs on the brain, and so on. This was an important finding that changed an entire branch of behavioral research that relied on water mazes," says Boleslaw Srebro.

The temperature effect was an important finding in itself, but it also meant that the field potential method was ill-suited to providing any information about learning. Still, Edvard didn't abandon hope of seeing a learning effect. His solution was to conduct two different experiments at the same temperature: one in which the rats swam around and learned, and a control experiment in which they simply sat still in a bucket of water at the same temperature as the water maze. When he analyzed the recordings from the control experiment, he saw a less prominent effect that had been hidden by the much larger temperature effect. Nevertheless, his conclusion was that they would have to abandon the field potential recordings. There were far too many factors that influenced them.

There was, however, another possibility: single-cell recordings. This would require him to use electrodes around five times smaller than those he had used before but would allow him to capture recordings from a single neuron instead of large groups of cells. Two main research groups had perfected this technique – John O'Keefe's group in London, and Carol Barnes and Bruce McNaughton's in the US. To learn the technique, Edvard would have to spend time with one of these groups.

Meanwhile, May-Britt followed her original plan. She trained rats, then put them down and cut their brains into slices. Finally, she spent hours, days, and weeks in the dark little room with the confocal microscope, counting the synapses from the neurons of rats that had lived in stimulating cages and those that had lived in normal rat cages. She didn't know herself which group the rats belonged to as she counted. When she finished, she had proved that animals that had grown up in stimulating cages had many more synapses than the ones that had spent their entire lives in less stimulating cages. It was in fact possible to see traces of learning in the hippocampus. The findings were published in the prestigious journal, *Proceedings of the National Academy of Sciences of the United States* (PNAS).[21]

In fall 1994, all six PhD students were approaching the end of their grant period. That was when May-Britt and Edvard hit on an idea that they proposed to the others. What if they all defended their dissertations at the same time and had a joint graduation? The idea was that they could help one another along if they were all heading for the finish line at the same time. Their fellow students immediately warmed to the idea, but Andersen was skeptical. He was afraid that one of them could delay the work of the rest if anything went wrong. But when he saw how enthusiastic they were, he gave way. Soon, he was as enthusiastic as his students and started to refer to the event as the "Grand Slam" – a term borrowed from the world of tennis that is applied to players who win all four major tournaments. It occurred to him that this could be an opportunity to bring groups of international scientists to Oslo at the same time. Six whole doctoral defenses in a short period would require a lot of opponents. This provided an opportunity to organize a scientific symposium revolving around the topic of the hippocampus.

Although everybody worked flat out through the fall and spring, none of them came even close to meeting the original deadline of June 1, 1995. May-Britt and Edvard were expecting their second child in early June, so it was critical that they complete their research before they had entirely different business to attend to – but there was too much work outstanding. On Thursday, June, 2 May-Britt sat deep in concentration at her computer, writing an article related to her dissertation. As she sat there, she detected the first signs of labor. But she continued working; she didn't want to stop until she had finished.

"It's this kind of burning idea of simply getting something done, a very powerful motivation to figure things out, no matter when or where in the world. I was the same myself, so I understood her. If I'd been worried from a medical standpoint, I'd have put my foot down, but I wasn't. The hospital wasn't far away," Edvard Moser says.

Still, May-Britt didn't manage to finish her work that evening. The next morning, their second daughter, Ailin Marlene, was born. On Monday, May-Britt returned to the department to show off her daughter and continue working on her article. But by now some of the professors thought it was time to intervene.

"I remember that Per came to me and said: 'You have to go home with the baby.' And I was so disappointed that he said that because I knew he actually thought it was okay. It was my decision and he could see both the child and I were healthy," May-Britt Moser says.

Three months after the deadline, all of them had finished their dissertations. In December 1995, all six scholars defended their dissertations over the course of a week. This event was so noteworthy that the Norwegian daily *Aftenposten* dedicated two whole pages to it. Beneath the headline "Use Your Head," was a photo of Andersen surrounded by four of his PhD students. The article described the scholars as "disciplined and coordinated troops who had conducted a veritable military campaign to conquer the hippocampus under the leadership of Per Andersen."[22]

The defense also gave Edvard, May-Britt, and the other students opportunities to network with international scientists such as the American couple, Carol Barnes and Bruce McNaughton, and John O'Keefe, who was based in London.

Carol Barnes, the opponent for May-Britt's defense, laughs when she looks back on it. Although she had been an opponent many times before, she had never been involved in anything like this.

"May-Britt is the most unusual person. I am asking her tough questions and she is doing well, of course, but then she starts asking the opponent questions back.

And I never had a student, somebody whom I am questioning, have questions for me as well. And we sort of laugh about it," Barnes says.

Over the course of the following week, doctoral defenses took place during the day, and sleigh trips and dinners in the evenings. On December 13, the PhD students even organized a St Lucy's day parade with their children in honor of the foreign guests. On the last day, when Edvard defended his thesis, events concluded with a huge dinner for more than 100 guests.

"The performance of all the students on that day was thrilling. It has to have been one of pinnacles of Per's life, to have those outstanding young scientists coming from his lab. He was like a proud father. It was just a spectacular event," Barnes says.

It was time to move on, learn from others, and launch careers as independent scientists. May-Britt and Edvard went to Edinburgh with their daughters Isabel and Ailin. Together with Richard Morris, they would make a fresh attempt to link LTP and learning.

CHAPTER 7

Edinburgh

Driving into Edinburgh for the first time, it's impossible not to be impressed by Edinburgh Castle. It looms majestically over the city, half protective, half threatening, from its platform of volcanic rock. Norwegian visitors to the city quickly notice the absence of wooden houses; everything is built of stone. The city center also bears the stamp of the university that has been there since 1538. Imposing university buildings are scattered throughout the old town. The university has many eminent figures among its alumni, including a young Charles Darwin, who studied medicine at Edinburgh before becoming an evolutionary biologist.

Apart from some obvious differences, May-Britt and Edvard found the place reminiscent of their childhood home in Sunnmøre. The city lay on a fjord – the Firth of Forth – with islets and skerries, and the weather was often the same as at home, wet and windswept. Richard Morris, who was based there, was one of the true greats of memory research. Since meeting him for the first time by their poster at the conference in Stockholm, the Mosers had kept in touch with him. They had even been on several short research trips to visit him.

In spring 1996, they returned as post-docs. Now their plan was to take a closer look at LTP. Several research groups had attempted to determine whether LTP or LTP-like processes were involved in animal learning, but the answers had been inconclusive. Many had tried to produce a kind of artificial saturation of the brain's capacity for learning by sending electrical impulses into the hippocampus. If it was true that LTP was important, this sort of saturation should prevent the animals from learning. Some found this to be the case, while others found that, on the contrary, learning increased. In other words, confusion continued to reign in the field. Together with Morris, May-Britt and Edvard hoped to settle the debate once and for all.

They had a few ideas. First of all, they would render the hippocampus in one half of the brain inactive using ibotenic acid so that they would have a smaller area to explore. In addition, they suspected that earlier attempts at saturating the neurons with impulses might not have been thorough enough. In order to reach as many neurons as possible, they planned to set up an array of electrodes at both inputs of the hippocampus.

At that time, Morris's research group occupied an entire floor of Appleton Tower, a modernist building from the 1960s that loomed squarely above the 18th century buildings around George Square in the old town. Here, Morris had one of the most spectacular offices in the entire city. His seventh-floor windows offered an unbelievable view of the castle and the northern part of the city. The lab facilities where May-Britt and Edvard worked were on the floor above. They were bright and spacious, and the windows overlooked the southern part of the city. To the east they had a glimpse of Holyrood Park and Edinburgh's mossy green mountain, Arthur's Seat – the highest in the city, although only 250 meters in altitude. From the large windows they could watch the rain, wind, and sun taking turns to gain the upper hand over the city.

Apart from a few short walks, the view from their laboratory windows was almost all they saw of Edinburgh. They worked pretty much all the time. Their girls had places at a daycare center in the city, but after closing time, they often joined the couple at the lab, where they had their own play corner. Only one time during their stay did the family take a short holiday, traveling to the west coast of Scotland, north of Glasgow.

Richard Morris, an Englishman, had come to Scotland after completing his PhD and gotten his first job at St. Andrew's University, just north of Edinburgh.

It was there that he had come up with the idea for the maze that bears his name.[23] The laboratory where he worked was down by the fjord, in the same building as a marine biology station. Every day on his way to work, he walked past a series of large water tanks containing fish, crabs, and all kinds of marine animals. One day while he was walking past, it occurred to him that he could use one of these tanks for his rat research. If there was a platform hidden beneath the surface of the water, a rat could neither see nor smell it before coming into contact with it, and that meant you could be certain it had found its way solely by use of its memory. It turned out to work extremely well. Aided by the water maze, Morris proved that rats that were missing part of their hippocampus were much worse at finding their way to this platform than control rats, confirming that the hippocampus was vital for remembering places. In his most important work, he showed that if one blocked receptors in rats' brains that were associated with LTP, they couldn't learn where the platform was either.[24] If Edvard and May-Britt could show that the rats were incapable of learning when their neural pathways were saturated, this would be conclusive proof that LTP was important for learning.

This was easier said than done. The experiments didn't work that well, and the work was frustrating. But they had enough research experience to know that this was normal. If you just held out long enough and didn't give up, and perhaps altered your course a little, things would work out one way or another.

Some months into their time in Edinburgh something happened that further delayed their project. A surprising job offer suddenly came up in their homeland. Although they were inclined to spend more time abroad, they had applied for an Assistant Professorship in the psychology department of the Norwegian University of Science and Technology in Trondheim (NTNU). They had given themselves the luxury of making some pretty unprecedented demands. Among other things, they had delivered a list of equipment they would need and made it clear that they weren't interested in moving if only one of them got a job. It was either both of them or neither. They were pretty surprised when the university acceded to all their demands and offered them two posts. It was almost too good to be true.

They didn't know what to do, so they contacted several colleagues to ask for advice, including Bolek Srebro, and Carol Barnes and Bruce MacNaughton. Everybody advised them to accept, reminding them how rare it was for both members of a couple to be offered posts in the same department. They decided to take the jobs. That meant they would have to be there for the start of the term in August. This created a dilemma. After discovering that brain temperature could influence the recordings, Edvard and May-Britt had become convinced that single-cell recordings were the way forward. If you recorded from several cells at a time – the technique they were currently using – many factors could influence the recordings. Edvard and May-Britt didn't want to go back to Norway before mastering the technique for recording from individual cells.

Morris suggested that Edvard spend the last three months of his post-doc with John O'Keefe, with Morris continuing to finance his post. So Edvard went on ahead to London while May-Britt remained temporarily in Edinburgh, where the girls had places in daycare.

The next stop was Gower Street in London, home to the world's foremost community for research into the sense of place.

CHAPTER 8

An Inner Map

From a bird's eye view, London looks like a huge, complex maze. A patchwork of boroughs, broad and narrow streets, buildings, and railroad lines, dotted with the occasional patch of green. It's a wonder the people of this megacity can find their way around.

It was unusually hot and humid in the English capital as May-Britt and Edvard walked down Gower Street, Bloomsbury, early one morning in June 1996, turning in through the great oak doors of the Anatomy Building and hurrying up the stairs with their wrought-iron and oak banisters to the top floor. The staircase ended in a dead end, a narrow door standing alone on the topmost step. But the door concealed something that was of great scientific value to Edvard and May-Britt. Behind it, in a cramped little laboratory, John O'Keefe had spent many decades pioneering the development of a method for recording from single cells in the hippocampus. Edvard and May-Britt had first met O'Keefe when he was an opponent during their doctoral defense in Oslo. The two PhD students had made an impression on him.

"I am really good at spotting good students. So when they approached me and asked if they could come and learn single unit recordings in my lab, I was very, very eager to have them. I hoped they would go back to Norway and set up their own lab," O'Keefe says.[25]

Like Edvard and May-Britt, New Yorker John O'Keefe was originally interested in memory and how it was formed in the hippocampus. He had come to University College London in the late sixties with a doctorate in psychology from McGill University in Canada. He had a diverse background, having worked as an aeronautical engineer and having pursued studies in film-making, English literature, and philosophy.[26] One of O'Keefe's teachers at McGill was the psychologist Brenda Milner, who had studied the famous memory patient, H.M. Like so many others, O'Keefe was greatly inspired by Milner's findings, and he was convinced she was correct in believing the hippocampus to be the center of episodic memory.[27]

But how was memory formed in the hippocampus? Would it be possible to see traces of memory in individual cells? O'Keefe was lucky enough to start his research career at precisely the time when technological developments made it possible to seek answers. In Canada he had developed the equipment and techniques that would enable him to make recordings of the electrical spikes from neurons. Electronic amplifiers that were small enough to be attached to electrodes on the heads of lab animals made it much easier to study the animals while they were awake and moving around. The recording equipment was upgraded with portable amplifiers and thinner wires, which made it possible for the animals to move freely.

The first brain structure O'Keefe studied using this method was the amygdala in cats. The amygdala is often referred to as the brain's fear center, but the structure is also important for a larger spectrum of feelings. O'Keefe realized that cells that were mostly inactive were often worth monitoring extra carefully. While the cells that were active all the time were difficult to link to a specific event, it was much easier to understand what a silent cell was reacting to in the environment when it suddenly fired off an electrical impulse. By monitoring the silent cells, O'Keefe found cells in the cat's amygdala that reacted to mice (mouse-detecting cells), others that responded to birdsong, and yet others that fired when the cat was presented with certain foods. This experience prompted O'Keefe to develop his own "natural law" of the brain: that the silent cells are the most interesting ones.[28]

One day in 1970, O'Keefe was using the technique to study cells in the thalamus, when he missed his mark and hit the hippocampus instead. Here he encountered a cell whose activity was very clearly linked to the animal's movement – a peculiar finding in a brain structure that was, according to established science, linked to learning and memory.[29] This was so interesting that he decided to drop his original project and concentrate on the hippocampus instead. Together with a master's student, Jonathan Dostrovsky, he started to record activity in the hippocampal cells of rats while the animals engaged in various behaviors such as eating, grooming, and exploring their surroundings.

O'Keefe and Dostrovsky noticed two different types of cells: some that fired often and seemed to be correlated with the animal's movements and others that were mostly silent, but would fire at irregular intervals. Based on his experience with the silent cells in the amygdala, O'Keefe was particularly interested in the latter, but initially it seemed impossible to find any link between the activity in these cells and the animal's behavior.

The turning point came one day when they obtained particularly good signals from a cell. At first it was absolutely silent. Then they heard a pop-pop-pop-pop-pop sound as the rat wandered into a corner of the box. Then it fell silent again when the rat moved to another place. And then came another pop-pop-pop-pop when the rat later returned to the same corner. Suddenly, in a classic Eureka moment, it dawned on O'Keefe that it wasn't *what* the animal was doing or *why* it was doing it that triggered the activity in the cell. The cell only fired at a particular place in the experimental box. This cell was a place detector! To confirm the results, O'Keefe and Dostrovsky immediately altered the surroundings to see if that influenced the activity of the cell from which they were recording. The cell wasn't affected by small changes, but if they made large changes – like drawing aside the curtains that surrounded the recording box – they could abruptly alter the cell's activity. The changes in the environment had to be so significant that the rat no longer perceived it as the same place.

In the days that followed, it occurred to O'Keefe that his finding could be evidence of the cognitive map that the American psychologist Edvard Tolman had described in his article "Cognitive Maps in Rats and Men" in the late 1940s.[30]

Tolman had trained rats to search for rewards that were hidden in a maze. After the rats had spent a few days perfecting the route to the reward, Tolman altered the maze such that one of the usual routes was blocked. But the rats weren't fazed by their encounter with the blocked corridor. They seemed to know which direction to take and chose another corridor that led them to the reward. In a series of cleverly designed experiments, Tolman and his colleagues altered the original maze in different ways. In each experiment, most rats chose the route that led in the right direction. It was difficult for Tolman to draw any conclusion other than that the rats were forming a mental map that provided them with an overview of the maze as they explored it. Despite the conclusiveness of his findings, they failed to convince the psychological research community at the time. As a result, the theory never took off, and the idea of the cognitive map vanished into oblivion in the decades that followed.

Could it be this map that O'Keefe had found in the hippocampus?

O'Keefe co-wrote an article with Dostrovsky about the cells they had found, in which they suggested that this was the cognitive map described by Tolman.[31] They expected a storm of reactions, but, with few exceptions, almost no one in the field seemed interested in the finding.

Among those who did show interest was American neuroscientist James B. Ranck Jr., who eventually replicated O'Keefe's finding. In 1984, Ranck found an additional type of map cell: head-direction cells, which indicated to the rat the direction its head was facing. The cells were located in a brain structure adjacent to the hippocampus called the presubiculum. This appeared to offer further support for the cognitive map theory.

One factor that proved especially important in disseminating the idea of the cognitive map was O'Keefe's collaboration with the neuropsychologist Lynn Nadel. The two developed a theory about how the cognitive map might work, and soon they sent a 50-page manuscript around to colleagues for review and comments. It took several years before the manuscript was ready to be published as the book entitled *The Hippocampus as a Cognitive Map*,[32] and in the meantime the text gained status as a kind of underground Bible in the research community. Many scientists understood how useful these cells that encoded position, which they called place cells, could be for understanding other psychological phenomena such as memory and learning.

"The book really gave a foundation of animal work in understanding brain function in humans. The kinds of test that you could do in animals to tap into the hippocampus seemed to be those that could cross species. It doesn't necessarily require language, if there is a spatial component, the rat doesn't have to *tell* you I know where I am, he will show you that," Carol Barnes says.[33]

"The book was almost a revelation. And it still holds up well," says Boleslaw Srebro, one of the few scientists in Norway to receive the manuscript for review.[34]

By 1996, a great deal was known about how place cells behaved. Scientists knew that they were located in the CA1 and CA3 regions of the hippocampus and that they fired in a limited area within an environment called a place field (Figure 4a). But there were still many more questions than answers when it came to how animals used the system to orient themselves. On the top floor of Gower

Street in London, O'Keefe introduced Edvard and May-Britt to everything he knew about place cells and the technique for recording from them. He showed them how to surgically implant thin electrodes, screwing them slowly but surely down into the hippocampus, while listening to the electrical signals they picked up until they heard the characteristic sounds of the place cells. Edvard and May-Britt were delighted to find their first place cells and hear the popping sound when the rat came to a particular place.

Space was so tight in the O'Keefe group that Edvard and May-Britt had to share a chair and a computer with a third researcher, Neil Burgess. They managed to make it work by taking turns being in the laboratory and taking lunch breaks. The cramped conditions also meant that O'Keefe was always nearby so they could ask him questions. For both Edvard and May-Britt, the weeks spent with O'Keefe were among the most instructive of their lives. He showed them not only how to record from single neurons, but also how to analyze the data. May-Britt and O'Keefe sat together for eight hours at a time as he gave her a step-by-step demonstration of how to operate on the rats. She noted his cleanliness and precision as well as his respect for the lab animals.

The more they learned, the more they were amazed by these strange place signals. How did these cells deep inside a rat's brain know the animal's location?

CHAPTER 9
A Dogma Overturned

As students in Oslo, Edvard and May-Britt often daydreamed together about the department they would someday build. But never would they have believed they would have free rein to do whatever they wanted so soon after completing their PhDs.

"Being allowed to decide for ourselves what we wanted to focus on, how we wanted to do things, where we would use our resources – it was just such a big deal. I felt like a young foal let out to pasture. Oh, it was such fun!" May-Britt Moser recalls.

"It wasn't scary at all. We'd already set up our own lab three times before, so we had experience doing it," Edvard Moser adds.

From the moment they started at the Norwegian University of Science and Technology (NTNU) in Trondheim, in autumn 1996, it took around a year to install all the equipment in the converted bomb shelter that they had been assigned as a laboratory. First they had to build a pen for the rats and an operating room, as well as several rooms for animal testing – including water maze experiments. Then, in addition to teaching psychology students, they had to tend and train the rats. So when at last they got to hear the popping sounds of a place cell in their own laboratory, they were keen to share the experience with as many people as possible.

The head of the psychology department, Sturla Krekling, was happy for them; he knew that this called for celebration. The popping sounds were evidence that the equipment was working. Even the cleaning staff and janitor stopped by to hear the sound of these cells indicating to the rats precisely where they were. They knew from the joy that May-Britt and Edvard radiated that this was a huge accomplishment.

While building a laboratory for single-cell recordings, the Mosers also resumed the project that they had worked on with Richard Morris in Edinburgh, this time beginning from scratch. Months of intensive work followed. First they deactivated one side of the hippocampus in 43 lab rats by injecting the brain with ibotenic acid, which killed nerve cells in the surrounding area. Then, after a two-week recovery period, they surgically implanted electrodes in the rats' brains. May-Britt and Edvard sat by the cage of each rat as it woke up from the anesthetic to monitor its recovery. Two weeks later, the rats were placed in a familiar dark box in which they felt safe, and their brains were stimulated with electrical impulses via the electrodes to "saturate" their neural pathways. Finally, they were ready to perform the conclusive "test" of the rats' memories. One by one, the rats were carefully released into water heated to body temperature, where they immediately began to swim around, searching for solid ground beneath their feet. Once a rat found the platform, it was removed and allowed to rest before being released back into the water.

As predicted, rats in the control group, whose neural pathways had not been "saturated," learned to swim straight to the hidden platform. But what about the other rats, whose neural pathways had been saturated? The Mosers hoped that their neurons would be unable to undergo long-term potentiation (LTP) and that those rats would prove to be worse than the control group at finding their way back to the platform.

The experiments were long and exhausting, and their working days sometimes lasted nearly 18 hours. To accommodate family life, Edvard and May-Britt worked shifts. They were convinced that it was critical to complete all the training in a single day. If it was drawn out over several weeks, the rats' brains might begin to compensate for any changes induced by the electrical stimulation.

The results confirmed that the rats whose neural pathways had been saturated had greater difficulty learning where the platform was than the control rats. In all, this suggested that LTP was in fact important for learning. These findings resulted in the first *Science* article produced by the Mosers' own lab in Trondheim.[35]

Even so, it wasn't all plain sailing. The Mosers felt isolated in the psychology department. Unlike the rest of their colleagues, they weren't doing research on human beings but on rats. There was little interest in the biological basis of psychology. One exception was Sturla Krekling, who himself had some experience with animal research. But by and large, it was clear that the department wasn't prioritizing the academic direction that May-Britt and Edvard were taking. The Mosers even published in different journals from their colleagues.

"We've always been oddballs and a bit out of the ordinary. When we were psychology students, we used to head to the lab and run experiments late into the night instead of going to parties to drink red wine and discuss terminology. In Per Andersen's lab we weren't medical practitioners like the rest of them but psychologists. In the psychology department, we were seen as medical practitioners. But there's a freedom in oddity, too, because you can do what interests you," May-Britt Moser says.

So the Mosers turned to other academic fields like biology and physics in search of collaborators. Rather unexpectedly, the janitor and cleaning staff also became important supporters, playing a decisive role in the way they pursued their lab work.

For May-Britt and Edvard, the rats' wellbeing had always been a priority. Yet they were well aware that others might not approve of their use of animals. In the United Kingdom, where they had worked the previous year, animal research was kept out of the public eye. It wasn't uncommon for militant animal welfare movements like the Animal Liberation Front to attack medical research facilities that engaged in animal research. While they were in the UK, they were warned not to tell the pre-school teachers at their children's daycare center that they worked on animal research because the daycare staff might have links to people who were willing to let off bombs.

"We were ridiculously scared," May-Britt Moser says of the atmosphere in the UK.

The Animal Liberation Front had offshoots all over the world, including in Norway, and was known for planning and executing its attacks with the assistance of people known as "hands" – insiders employed at research institutions.[36]

So when they were about to start work in Trondheim, May-Britt and Edvard became extremely nervous about the idea of the janitor and the cleaning staff having keys to the laboratory. They even discussed with the department's office manager, who had previously worked in the army, whether they needed to contact the police to map security risks and run security screenings on the employees. The more they thought of all the things that could go wrong, the more frightened and agitated they became.

But they soon realized how absurd all this was when they met the janitor, Svein Erik Simonsen, and the cleaning women.

"We went from being ultra paranoid to thinking: We're in Norway, we're in Trøndelag. Calm down! It was a very important experience because then and there we decided that we would run an open lab. Anyone who wants can come in and look around; we have nothing to hide. We're doing the best we can, we're as humane as possible. When the janitor and the cleaning staff are allowed in to look around and react to our findings with interest rather than indignation, it reassures us that everything's fine."

Indeed, they became fond of the janitor and the cleaning staff, considering them part of the team – especially Svein Erik Simonsen. Whatever they needed, he was always there.

Initially it was just Edvard and May-Britt working together. Since there were no technicians, they had to do all the work themselves, everything from taking care of the animals and cleaning their cages to operating on the rats and running experiments. In the early days, they didn't have any students either – most psychology students preferred patients to rats. Eventually, however, they managed to get a few more people on board.

The first one to join the lab was marine biologist Kyrre Haugen, an anarchist rocker type. He was employed as a technician, and May-Britt gave him stringent training in how to slice frozen rats' brains into perfect, microscopically thin slivers. To ensure that he earned a living wage, since they often didn't have the funding to pay him for months at a time, they also employed him privately. He would babysit the girls, and one summer he painted their house.

In 1998, they got their first PhD student, Stig Hollup – a burly man from Trøndelag who rode a motorbike and was extremely dexterous and skilled at building lab equipment. His main interest was biological psychology, and he had moved to Oslo to study neurobiology since there were no neurobiology courses in Trondheim. Naturally, Hollup was extremely interested when Edvard and May-Britt started up their lab in Trondheim.

Soon they were also joined by Vegard Brun, a young man from Tromsø who was completing his civilian national service. He had contacted them to find out if they needed anybody to assist with simple jobs like copying articles or to help out in the lab. They certainly did! At first, they set him to work cleaning cages and training rats. But soon they realized that he wasn't just talented but also motivated to do more. A few months later, he began to conduct his own experiments.

Eventually, they ended up with a small group of five or six students, all with very different backgrounds and personalities. Hill-Aina Steffenach, from northern Norway, had studied biology. She was an extrovert with thousands of irons in the fire. She was passionate about everything from traditional crafts to world music and knew people from every conceivable walk of life. The biologist Sturla Molden was skilled at programming and data analysis. Along with Mona Kolstø Otnæss, from Oslo, and Frode Tuvnes, they were like one big family. During intensive periods, the students often went back to Edvard and May-Britt's home in Lade to eat dinner before returning to the lab. And when they went to an overseas conference, they usually traveled together, the whole gang – including the technicians.

One of the first questions the Mosers and Hollup asked was whether the place cells' behavior was the same when the rats were swimming as when they were on dry land. This had to be determined before they could use the recordings from the water maze to draw any conclusions about the function of place cells. Nobody had recorded place cells in swimming rats before, and few people believed it was even possible, since the large water tank would act as an antenna for electrical interference.

Before they could get started, though, they had to resolve a practical problem – how to keep the cables away from the water while also allowing the rat to swim freely. At first, they developed a rather homespun solution: While the rat was swimming in the tank, one of them stood behind a folding screen and held the cables clear of the water with a fishing rod. However, the method had its drawbacks. For one, they had to take care to move the cables in the same direction the rat was swimming – a demanding task since they were standing behind a screen and could only follow the rat's movements on a monitor. It was also difficult to avoid, consciously or unconsciously, guiding the rat in a particular direction. Now and then, water would come into contact with the electrodes, and then there would be no recording.

Hollup learned to "drive the fishing rod," but at the same time, he began developing a solution that would make the rod superfluous and ensure that the researchers didn't lead the rats in a given direction. After much pondering, he came up with a system involving three pulleys and counterweights attached to the ceiling – this enabled the rat to swim freely without feeling the weight of the cables and without the cables ending up in the water. They also made the recording devices as watertight as possible by wrapping the electrodes in plastic and coating the exterior in Vaseline. The system was so watertight that the rat could dive under the water without any adverse effects.

After around a year's work, they became the first people in the world to record place cells in swimming rats. Hollup's experiments eventually showed that place cells behaved similarly in water and on land. This meant that they could use recordings from swimming rats to study place cells. To have achieved something other people had believed impossible felt like a victory. And soon, other victories would follow.

A few years after starting at NTNU, the Norwegian Research Council awarded them strategic research funds. The criterion for receiving this support was that the applicants had to fall within the universities' strategic priority areas. Edvard and May-Britt applied. Deans Eivind Hiis Hauge, Torbjørn Digernes, and Gunnar Bovim, each of whom in turn later went on to be chancellor of NTNU, had the job of reviewing the applications. All of them agreed to award the first place to a group of aluminum researchers whose work fell squarely within the priority area of NTNU, a science and technology university. The second place gave them more trouble. The Moser group received excellent feedback, but neuroscience wasn't named in NTNU's strategic plan. Eivind Hiis Hauge explains their thinking:

"The biology of memory – what does that have to do with NTNU? But all three of us looked at one another and said: strategy or no strategy, this is so good that we just have to award them the second place. One crucial quality for university heads on all levels is one: recognizing quality when you see it; and two: accepting the consequence of this realization and doing something about it. It's very important to make strategies, but you can't let yourself be straitjacketed by inside-the-box thinking," Eivind Hiis Hauge says.

The current chancellor, Gunnar Bovim, first met Edvard and May-Britt in around 1998.

"The thing that set them apart was that they invited themselves into the best research communities and they were focused. The conversations weren't about what they needed: They always started out positively, beginning every conversation by telling us what they had achieved. They published in journals like *Science* and *Nature*; they were bold enough to send articles to the top journals straight away rather than trying to place them in less prestigious journals first," Gunnar Bovim says.

In 2001, Edvard and May-Britt moved their lab to the neighborhood of Øya to join their collaborators at the Faculty of Medicine. Jan Dyrstad, the head of the Faculty of Social Sciences, knew he was losing the cream of the crop, but at the same time, he had no doubt that letting them leave was the right thing to do.[37] The Medical-Technical Research Center at Øya was more suited to animal research: There they had their own animal facilities and veterinarians. One disadvantage of the move, however, was that the Mosers lost the thick, sound-muffling walls of the former bomb shelter. The new laboratory was just a few hundred meters from the main road out of Trondheim, so the ground constantly shook from heavy trucks driving by. Moreover, there was a large cellphone antenna fixed to the building. The air was alive with all kinds of noise. Eventually, they realized they would have to give up the water maze recordings. It was simply too technically difficult.

But another issue had become ever more pressing: Where did the place cells get their information from?

"You know, it's zeitgeist that makes some questions more obvious than others," May-Britt Moser says.

"God knows where the questions come from, but the fact is that even people who haven't spoken to each other may be dealing with things in parallel without knowing about the other person. And it was a bit like that with the most difficult question we'd had when we were in London – one that a lot of other people also had: What makes a place cell fire like a place cell? What kind of information does it receive to make it fire that way?"

The prevailing assumption was that the trisynaptic circuit discovered by Per Andersen in the sixties was the main pathway for the signals in the hippocampus. According to Andersen's theory, the signals first entered the hippocampus from the part of the cerebral cortex called the entorhinal cortex, via the dentate region, before continuing through CA3 to the end station, CA1. So it was natural to conclude that the place cells in CA1 must receive their information from CA3.

But would it be possible to confirm this? To test this, Edvard and May-Britt decided that they would destroy the CA3 area, breaking the signal loop in the hippocampus, and see whether that affected the place cells. Although this sounds simple, it was in practice extremely difficult. A rat hippocampus is only around 1cm long, and destroying a tiny, specific part of this structure without touching any of the neighboring areas is far from simple.

By then, Vegard Brun had been in their group for a couple of years, though he was neither employed by nor studying under them. He was a medical student, but he spent his evenings, afternoons, and vacations working in the Moser lab, in part because he loved doing surgery. He had just finished a job when Edvard Moser asked if he might be interested in a new project. When Brun heard what the project involved, he suggested using precise knife incisions to cut the connections between CA3 and CA1. They decided to try it out.

Vegard Brun and Hill-Aina Steffenbach worked on developing the technique. They would have to make the surgical equipment themselves. Steffenach cut razor blades into narrow 2mm-wide strips and attached them to narrow tubes they had in the laboratory. During the operation itself, they had to make five incisions inside the 1cm-long hippocampus, and the incisions had to be at a 45-degree angle along a central line. Although it was challenging, Brun didn't consider this the most difficult aspect.

"The challenge was inserting the electrodes that would be used to record the impulses and eventually also a cannula – all this within a radius of approximately 2mm – without damaging the brain tissue. The cannula was needed to inject a fluorescent substance that would show whether the two areas were still connected," says Vegard Brun.[38]

The first difficulty they could foresee was that they might not fully interrupt the connections, and as a result, some cells in CA3 would continue to send signals to CA1. To forestall this kind of difficulty, they decided to inject a fluorescent substance that would be absorbed by the neurons and dye them, from the offshoots to the body of the cell. After each experiment, they could check to see if any connections between CA3 and CA1 remained intact by looking at cross-sections of the brain. They had decided upon this after consulting with the Dutch anatomist and hippocampus expert, Menno Witter. Edvard and May-Britt had first contacted him when they were working on their master's theses and needed to know why the top of the hippocampus in rats was more important for place recognition than the bottom. Witter had been unable to give them a conclusive answer but had sent them a nice letter. After setting up their lab in Trondheim, the Mosers resumed contact with Witter. Since 2000, he had collaborated closely with Edvard and May-Britt, making extended visits to their lab several times a year. This time they needed his expertise on the neural connections in the hippocampus.

The Mosers expected to observe severe disruption of the activity of place cells once the connections between CA3 and CA1 were cut. This meant that the rat would no longer recognize places it had already explored. Although it wouldn't be able to tell the researchers this, they didn't need it to: The lack of activity in the place cells would be enough to confirm this.

But the results proved to be quite different. The rats with lesions were just as capable as the control animals of determining their location. As with the rats in John O'Keefe's lab, the place cells popped and crackled when the rats explored particular areas of the box, sniffing around in search of food. Vegard repeated the experiments with more rats and got the same results. The rats had no difficulty finding their way. These were remarkable results, which contradicted what people had hitherto believed about the signal pathways in the hippocampus. That the place cells in CA1 worked just as well as before meant one of two things: Either place cells produced the signals themselves or they obtained important information from somewhere else.

It seemed unlikely to May-Britt and Edvard that the place cells could produce the signals themselves. The activity of CA1 cells was less coordinated than among cells in CA3, which were interconnected in large networks. The most probable explanation was that the CA1 cells formed direct connections with the entorhinal cortex. Until then, hippocampus experts like Per Andersen had considered those connections to be of less importance. Could they have been mistaken?

The Mosers' discovery that place cells functioned normally even when the main pathway through the hippocampus was cut was published in *Science* in 2002.[39] Everything now indicated that they would have to look outside the hippocampus for the solution to the place cell mystery. They would have to make recordings in the entorhinal cortex, an area even more inaccessible than the hippocampus.

"Up until around 2001, the entorhinal cortex seemed scary, but now we were motivated to get to work as quickly as possible," Edvard Moser says.

The only problem was that other scientists had already searched this area, to no avail. Would they be able to find anything where others had abandoned hope?

CHAPTER 10

Into the Unknown

Marianne Fyhn switched on the electric shaver and bent over the little rat lying before her on the table. The rat's eyes were open, but the animal was anesthetized. The white sheen in its eyes was the ointment that would prevent them from drying out during surgery.

Carefully holding the head steady with her left hand, Fyhn used her right hand to shave off a square of fur over the rat's cranium. When she was satisfied, she put down the shaver, brushed away all the fur clippings, and sterilized the bald patch with alcohol. After that, she carefully moved the rat from its position on the table to a warmed cushion on the adjustable operating table right beside her.

Her task was to insert electrodes into the entorhinal cortex, which lies folded in toward the hippocampus, right at the back of the rat brain, just beside a large blood vessel. Nobody had ever undertaken an operation like this before. Fyhn had worked with May-Britt Moser and Menno Witter to develop the procedure for the surgery, and she was confident that she knew how to do it. This rat was the first animal that would spend the rest of its life with electrodes in its entorhinal cortex.

Fyhn had worked in the lab for around a year. She had a master's degree in biology and had originally applied for a technical post they had advertised. But when Edvard and May-Britt interviewed her, they saw a researcher, not a technician. They asked if she wouldn't rather complete a PhD – and that's what happened. Early on, she proved to be to be skilled at both operating on and training animals. She had the surgical skill needed to carry out this demanding procedure.

Fyhn gently placed the rat's teeth on a little gap in a metal strip to keep the head stable on the operating table. Then she pulled a small funnel over its snout so that it would receive a continuous supply of anesthetic gas that would keep it sedated. A small swelling on its back revealed where she had given the rat the fluid infusion its small body needed to remain hydrated while it lay on the operating table.

Fyhn secured the rat's head in the stereotaxic frame, using coordinates that would help her place the electrodes in precisely the right spot. This was painstaking work, but it was extremely important to invest time in getting it right from the start rather than risk having to go back and do it again later. When she was at last satisfied with the position of the rat's head and had checked once more that the animal was comfortable and that the anesthetic gas was flowing normally, she picked up a sterile scalpel and made a small incision above the rat's eyes toward the nape of its neck. She drew aside the skin, revealing the cranium. Now she could see the seams on the cranium, the marks left after the different bone plates fused together when the rat was a baby. These patterns can be used as a frame of reference when calculating stereotaxic coordinates, making it possible to find the correct area of the brain. She knew precisely where she was going. The three-dimensional route to the target was planned in advance. She had already tested out the procedure on a couple of rats and studied slices of their brains. The slices had confirmed that she had the right coordinates.

Menno Witter had been central to discussions about where they should search. Witter knew that substantial hints of the entorhinal cortex's importance to hippocampal function had existed for nearly 100 years. In 1902 the Spanish neurologist, Santiago Ramón y Cajal, who had mapped the brain's anatomy in great detail at the end of the late 1800s, had claimed that the hippocampus was closely connected to another area in the cerebral cortex.[40] According to Ramón y Cajal, the connections were so dense and prominent that they were visible even without a microscope. Based on these massive connections, he believed that whatever this area of the brain did, the hippocampus must do the same thing. The area he was talking about was what later became known as the entorhinal cortex.

In the 1980s and 1990s, German researchers had found that the entorhinal cortex was one of the first areas affected in human beings with Alzheimer's disease – and one of the first symptoms of Alzheimer's disease is loss of place recognition. Some research groups therefore tried to conduct more focused studies of the entorhinal cortex area in animal brains. Though they found some diffuse place activity, they concluded that the entorhinal cortex did not appear to be particularly involved in processing information about location in animals.[41,42]

However, Witter interpreted these results differently.[43] Earlier recordings in the entorhinal cortex had been carried out in an area linked to the lowest part of the hippocampus. The place cells located there fire in a fairly large field within the environment, which makes them difficult to record in the laboratory. The place cells with the most distinct fields are located in the part of the hippocampus closest to the top of the brain. Witter believed it was important to record in the central area of the entorhinal cortex, which has connections to those place cells. This area was difficult to reach in surgery because the atlases of the rat brain didn't include a map with the correct perspective.

In the operating room, Fyhn screwed the stereotaxic arm, to which the operating equipment was attached, to the stereotaxic frame. She then attached a drill with a small dental bit to the arm, as well as the implant with the electrodes that she would insert inside the brain. Having found the correct location for drilling, she lowered the dental bit toward the cranium. She began to drill extremely carefully. It was crucial not to damage the membrane of the brain inside the one-millimeter-thick cranium.

After a short while, she raised the drill and examined the cranium. There was a slight depression. The drill had not yet penetrated it. She lowered the drill again and drilled a bit more. Then she raised the drill and examined the cranium once more. Along the way, she checked how the rat was doing, and whether the anesthetic gas was still flowing properly. When she had topped up the fluids to ensure that the rat wasn't short of water, she continued drilling, microns (0.001mm) at a time, until the drill had passed completely through the cranium and she could see the thick brain membrane, the dura mater, which protected the underlying brain tissue.

She took out the scalpel and made a tiny incision in the membrane. Then she lowered the implant, a minuscule electronic device with extremely thin electrodes just 20 microns in diameter. Five such electrodes could fit within the diameter of an average hair. The electrodes were around the same thickness as the cell body of a neuron and so they would not puncture the cells. Even so, they were close enough to make it possible to capture signals from individual cells when they reacted to something.

Fyhn inserted the implant at the top of the cerebral cortex. The electrodes were still a long way from the area she was interested in. They would be screwed down into it slowly but surely once the animal had come around from surgery. She screwed the implant firmly to the cranium, then sealed it with dental cement – the kind that dentists use when they are fitting dentures. When she was finished at last, many hours had passed since she had started out. She detached the operating equipment from the rat and carried it over to another room with dim lighting. Here the rat was placed in a recovery chamber, a small plastic cage with reddish walls. Fyhn sat beside it, monitoring its recovery. May-Britt had taught all the researchers never to leave an animal alone when it was waking up from an operation.

Over the next hour, the rat came round from its sedation. The first signs were that its breathing changed and then its limbs started to move a bit. Eventually, the rat got to its feet, initially a bit groggy and dizzy, but soon increasingly steady. At first it was also unused to the new weight of the "hood" fastened to its head. Once it was able to stand on its feet, Fyhn gave it soaked pellets and a little baby porridge. Soon it was fully conscious.

For several days after the operation, the rat was left in peace to recuperate. It was important to ensure that it recovered and wasn't suffering any post-surgical infection. Once everything proved to be in order, Fyhn started training the rat to run around a small box on the floor. By nature, rats aren't keen on open landscapes, where they can easily fall prey to predators. So at first the rat hovered close to the edges. Fyhn placed pieces of chocolate cookie on the floor in the middle of the box to give it a reason to override its instinctive fear. Only after a few days' training, when the animal had become fully comfortable in the box, did it approach the center to retrieve the treat. Soon the rat had learned that there was nothing dangerous about venturing toward the center of the box, and it got increasingly better at gathering up the bits of cookie. Only once it had learned to crisscross the entire surface area of the box was it time to lower the electrodes down toward the entorhinal cortex.

Usually, when one turns a screw one rotation, it descends several millimeters. The screws for the electrodes worked differently. One full rotation caused the screw to descend just 200 microns. The electrodes had to be screwed down in minuscule steps to avoid lowering them too far. It's a matter of more haste less speed, as May-Britt used to remind her students. While the rat was training in the box, the electrodes were slowly screwed down through the cerebral cortex and deeper into the brain, 50 microns at a time. Since the brain is not equipped with pain receptors, the rat didn't notice a thing. Fyhn had some landmarks to guide her: neurons in the entorhinal cortex fire at a frequency of 8 hertz, for example. In order to hear whether she was close to any interesting cells, she listened to the cells as she proceeded, via speakers attached to the electrodes in the rat's brain. Meanwhile, the rat sat in a flowerpot, sniffing the room around it.

This painstaking process of screwing in the electrodes and training took several weeks, but one day at last they picked up signals from brain cells in the correct area, and it was time to start the recordings. The rat's task was always the same: to look for pieces of a chocolate cookie; but the big question for the researchers was how the cells in the entorhinal cortex would behave while the rat was racing around in the box. Would these cells also produce place fields like the place cells in the hippocampus?

Through the speaker, Fyhn could hear the sound of one cell firing while the rat ran around. It seemed to be firing at several places in the box. This was pretty unusual for place cells in the hippocampus, which generally became active in one particular place.

Once the data was analyzed, the whole team – she, May-Britt, Edvard, Sturla Molden, and Menno Witter – could see the activity patterns of individual neurons on the PC. Sturla Molden had developed a program that enabled them to correlate the rat's location in the box with the activity of its neurons. The cells were clearly responding to the animal's position, but not in the same way as place cells in the hippocampus. Whereas hippocampal place cells only produced an active field in one area of the box, the cells in the entorhinal cortex had active fields at several different places in the box.

Marianne Fyhn and May-Britt Moser followed up with several more rats and found similar activity. Now they knew that they had found the correct area. This was confirmation that the entorhinal cortex was significant for place recognition. It was also clear that the cells they had found were not O'Keefe's place cells. But exactly how did they differ from place cells?

To investigate the accuracy with which these cells indicated place, they needed a computer program to see whether the cell activity could be used to calculate where in the box an animal had been when a cell fired. Sturla Molden set to work on this. Although he had no formal training in mathematics, he was quick to grasp the mathematical aspects of a problem. He had what Edvard Moser called a math brain. His sharp programming skills enabled him to translate his mathematical insight into software that could analyze the recordings. Molden went through all the neurons for which they had recordings and discovered that the cells they had found in the entorhinal cortex were just as good as the place cells in the hippocampus at predicting where the animal was. Until then, the hippocampus had been viewed as central to mammals' inner cognitive map. Could the entorhinal cortex compete with the hippocampus for this label?

After an intense period of experiments, analysis, and discussions, the team were ready to make their results public, and in May 2004, they published yet another article in *Science*.[44]

But one issue in the article remained unresolved. There was something peculiar about the cells they had found, yet they couldn't put their finger on what it was. The neurons fired at several places in the same environment, and the firing pattern seemed strikingly regular. Fellow academics who read and reviewed the article noted that and encouraged them to go through their analyses again. As they reviewed the data once more, it became clear that the active areas of the cells were not spread across the environment randomly. But when they measured the distances between the points in search of a pattern, they came up blank.

"It was a great discovery, but it wasn't super revolutionary. People had seen place activity in the same brain area to a lesser degree. The mysterious thing was that each of these cells was active in several places. It was a strange pattern in the sense that there was a certain regularity in the fields, but we didn't understand what it was," Edvard Moser says.

What had they actually found?

May-Britt and Edvard on a tributary of the Amazon in Ecuador in 1986. Photo: private.

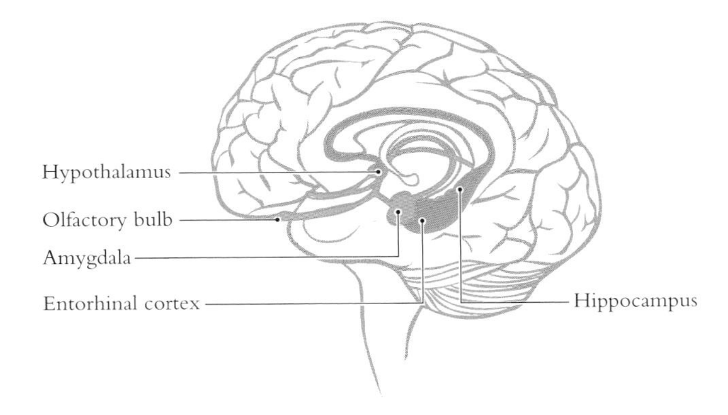

Figure 1. The hippocampus and its neighboring area, the entorhinal cortex, lie a few millimeters below the temples on either side of the brain. Experience with patients whose hippocampus had been removed suggested that this area was extremely important in the formation of memories of events. Illustration by Vigmostad & Bjørke.

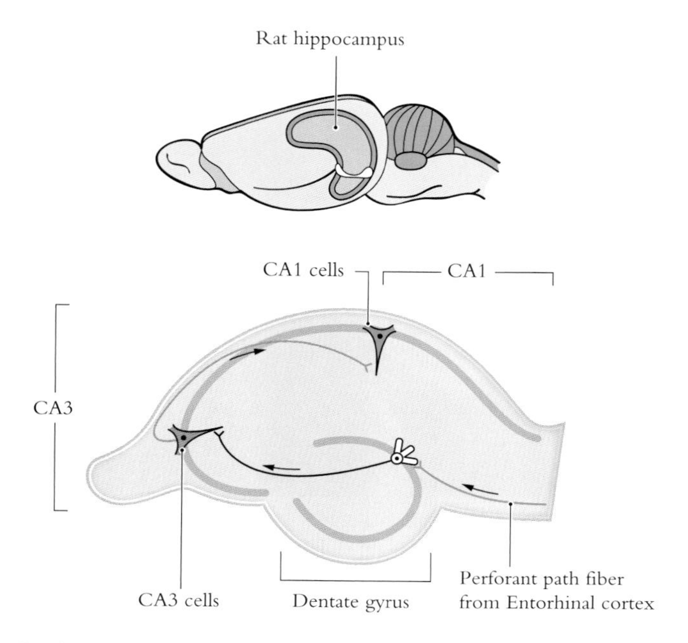

Figure 2. Inside the hippocampus, the signal traffic travels via the so-called trisynaptic circuit discovered by Per Anderson in the 1960s. Nerve signals enter the hippocampus through the perforant path before proceeding to the neurons in the dentate gyrus, then CA3, and finally CA1. Illustration by Vigmostad & Bjørke.

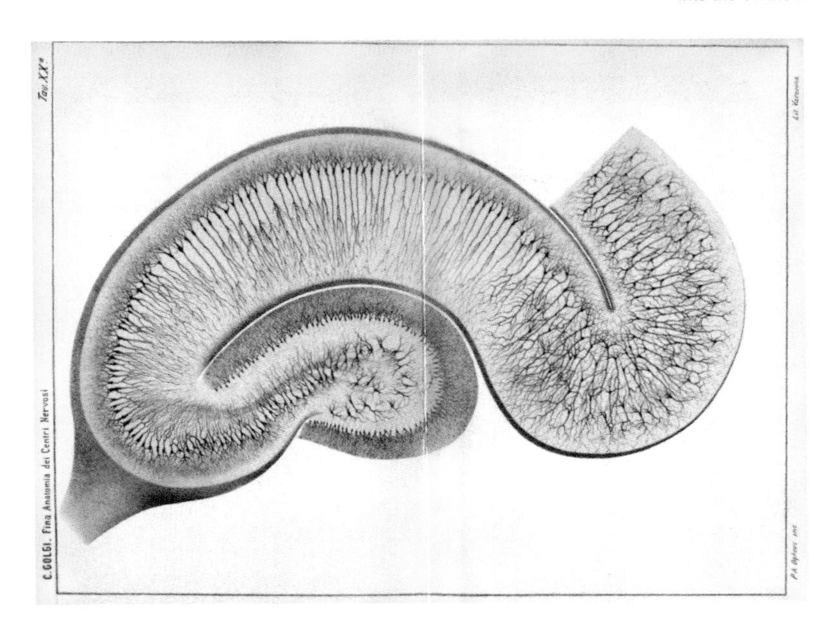

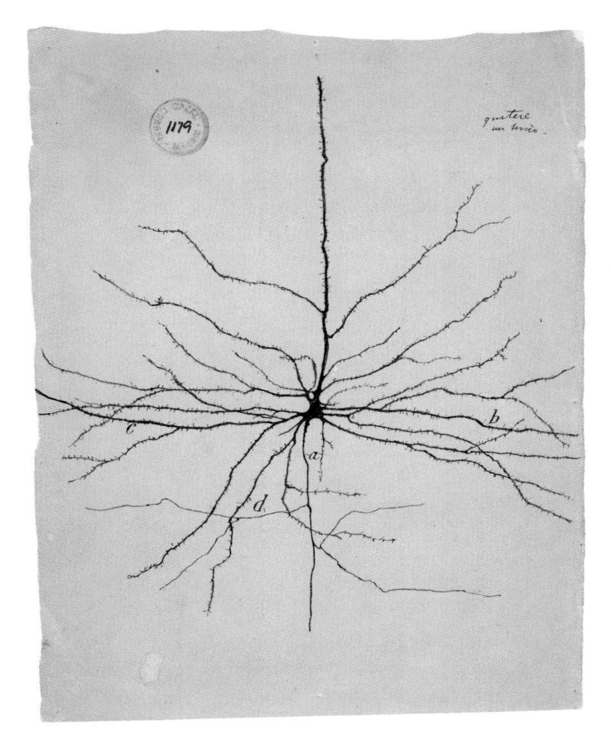

Above: From an early stage scientists were fascinated by the special shape of the hippocampus. The dyeing method developed by Camillo Golgi in the 1870s also revealed the beautiful, neat organization of the neurons in the hippocampus, as well as the massive connections to the rest of the brain.
Below: Sketch by Spanish scientist Santiago Ramón y Cajal, of a neuron in the cerebral cortex stained using the Golgi method.

SØNDAG

Søndag 10. desember 1995 Aftenposten

BRUK HODET!

Du har nok ant det men nå vet forskerne det: At det lønner seg å bruke hodet. Denne helgen skal hele seks norske doktorander utdype funnet.

KLOKE HODER: *Fire av de seks hippocampus-doktorandene og deres «troppsfører»: Fra venstre Mari Trommald, May-Britt Moser, Paola Pedarzani og Edvard Moser. I midten professor Per Andersen.*

In 1995, six of Per Andersen's students defended their dissertations simultaneously. All of them had done research on the hippocampus. Cutting from *Aftenposten*, 10 December 1995.

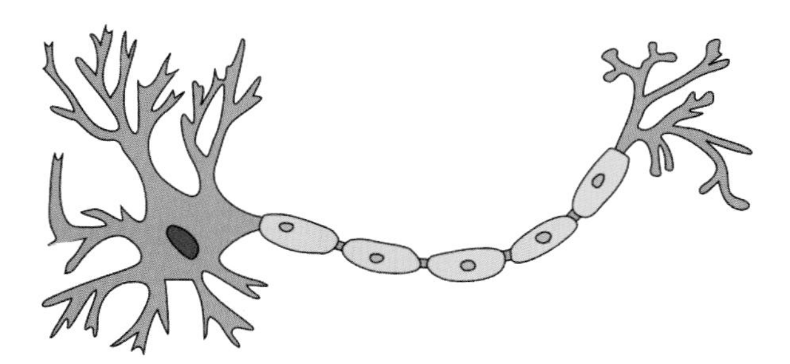

A neuron receives information via the dendrites, the fine little offshoots that extend from the cell body. Outgoing signals from the cells move along the axons in the form of an electrical impulse and are passed on to neighboring cells via the contact points, the synapses. Credit: Quasar (talk), Wikimedia commons, CC BY-SA 3.0

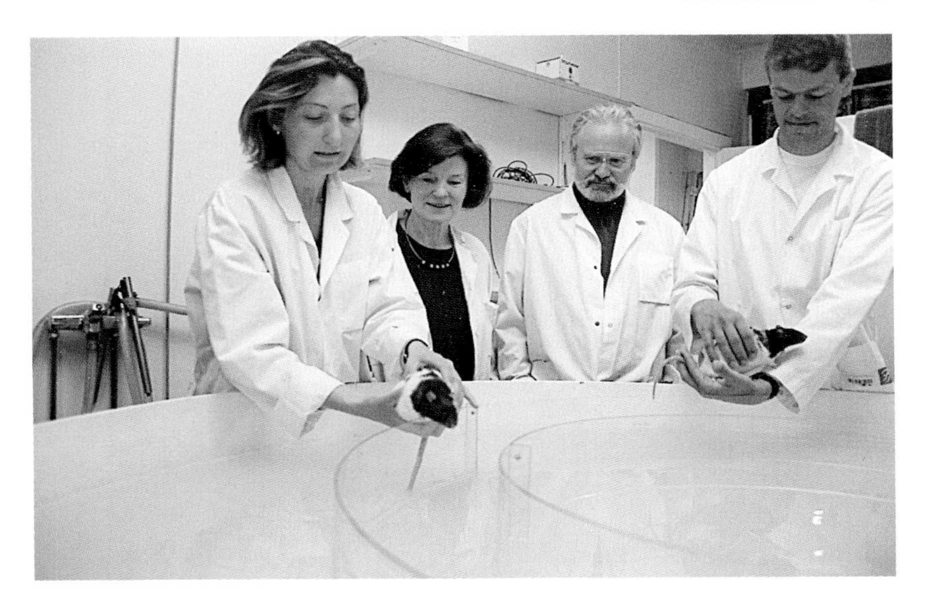

When May-Britt and Edvard started at NTNU in Trondheim, they embarked on a collaboration with natural science researchers in the absence of a biological psychology community. Here they are pictured by the water maze in the laboratory at Lade with biologist Hanna Mustarparta and physicist Arne Valberg. Photo: Lars Kr. Iversen.

The Moser group quickly gained a reputation for attending conferences with a whole wall of posters. Here, the research group is pictured in front of their wall of posters at a conference in Paris in 2002. From left: Edvard Moser, May-Britt Moser, Klaus Jenssen, Kirsten Kjelstrup, Vegard Brun, Mona Otnæss, Frode Tuvnes, and Ingvild Hammer. Photo: Ingvild Hammer.

Marianne Fyhn presents her research at a conference in San Diego in 2001. Photo: Ingvild Hammer.

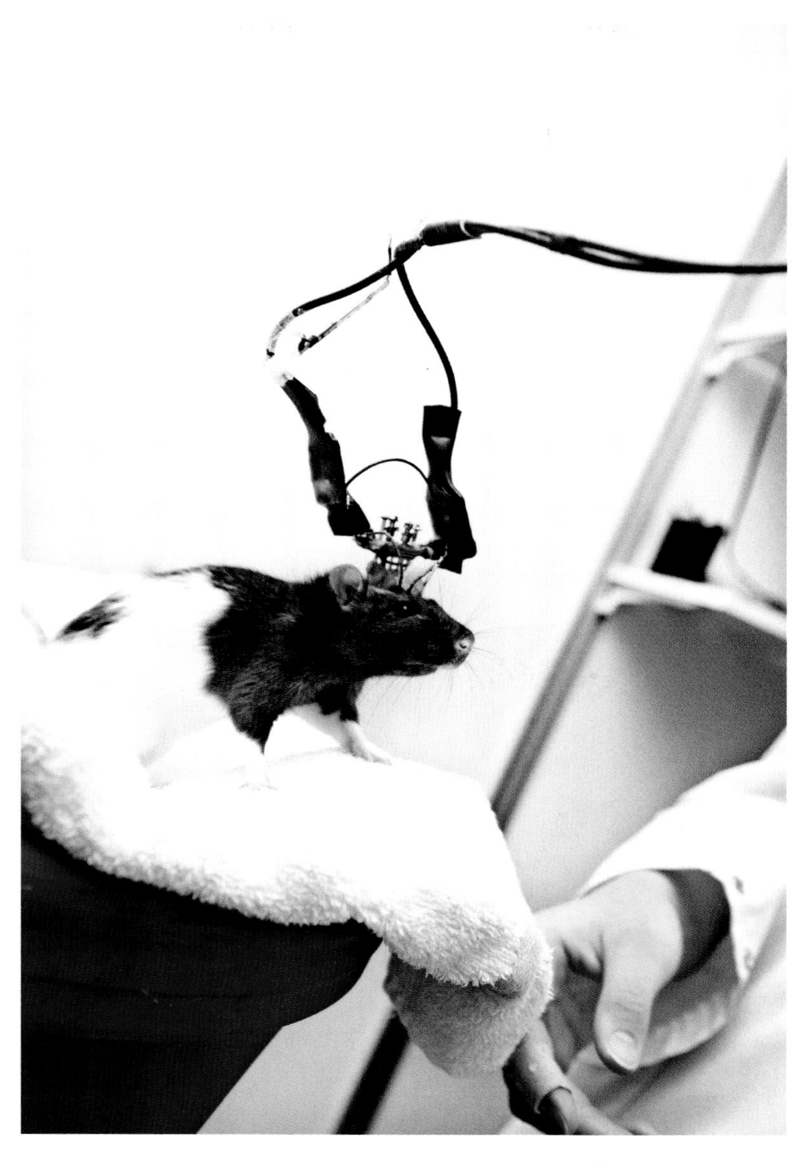

Between the experiments, the rat rests in a flowerpot as the researchers check that they are receiving good signals from the brain cells. Photo: Geir Mogen/Kavli Institute for Systems Neuroscience.

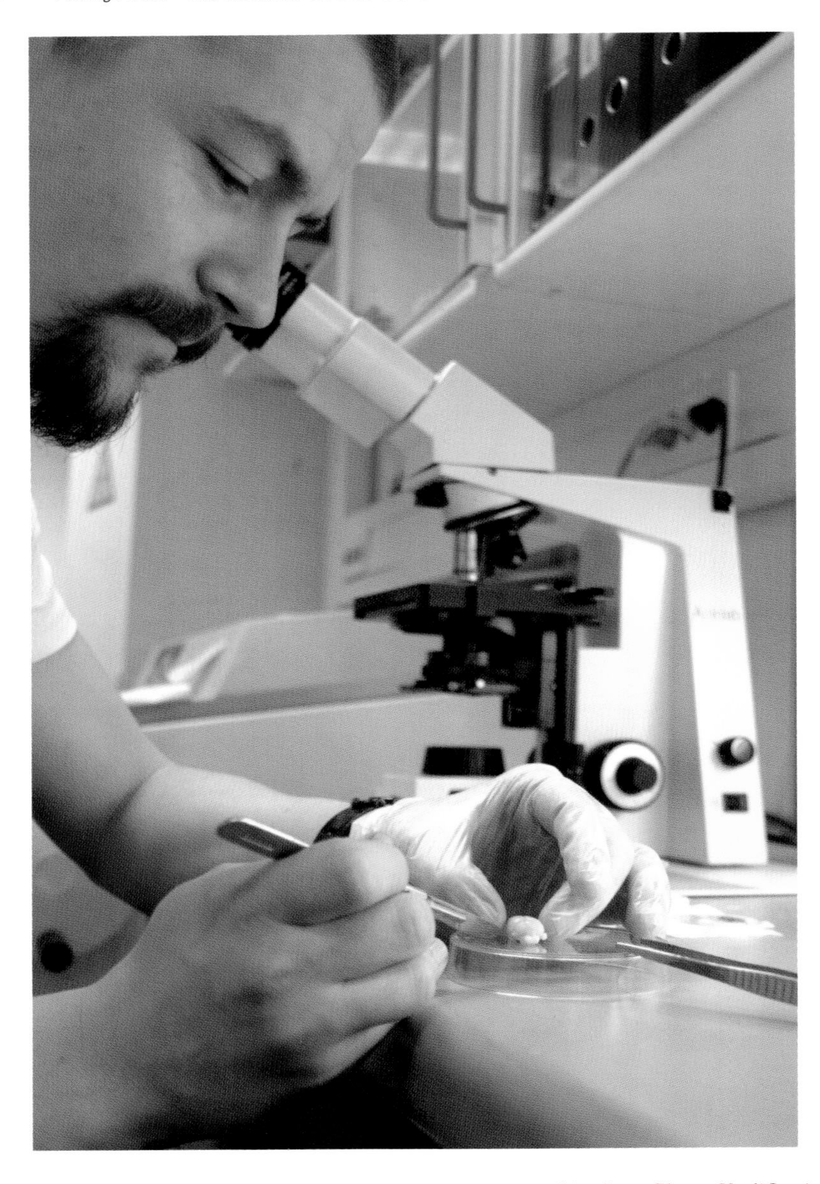

Kyrre Haugen with a rat brain he is preparing to cut into micro-thin slices. Photo: Kavli Institute for Systems Neuroscience.

After the experiments are done, the rats are put to sleep and their brains cut into micro-thin slices so that the researchers can document which tissue the recordings were made in. Photo: Bård Ivar Basmo

Vegard Brun with a rat on a linear track maze in 2006. Photo: Kavli Institute for Systems Neuroscience.

May-Britt Moser with a rat in a typical box that is used in experiments. Photo: Kavli Institute for Systems Neuroscience.

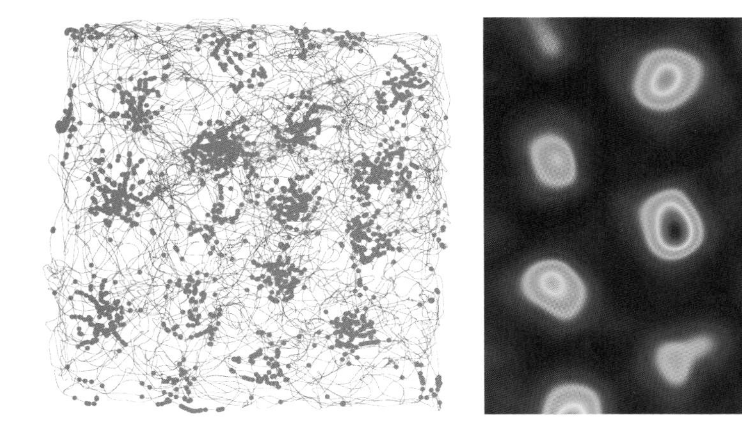

Figure 3. When Marianne Fyhn tested the rats in a larger box, they saw that the activity of the cells they were recording formed a hexagonal pattern. The lines in the left-hand figure show the rat's movements as it hunts for treats scattered on the floor. The blue spots show where in the box one cell in the entorhinal cortex reacts. In the right-hand figure, these data are run through software, and we see high activity represented by red and low activity by blue. The red dots form a hexagon. Photo: Kavli Institute for Systems Neuroscience.

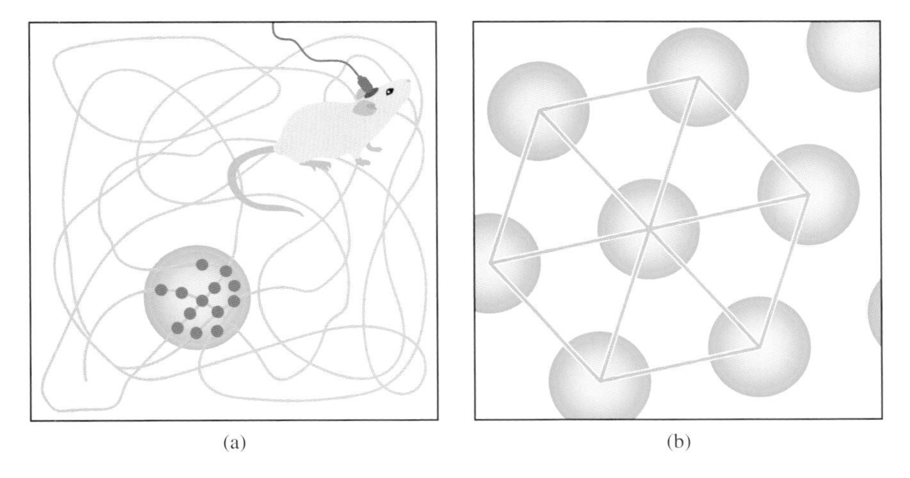

Figure 4, a and b. a) Place cells in the hippocampus typically fire at a specific place in an environment, a place field. b) The grid cells in the entorhinal cortex have many active fields per environment that form a pattern of equilateral triangles or hexagons. Illustration by Vigmostad & Bjørke.

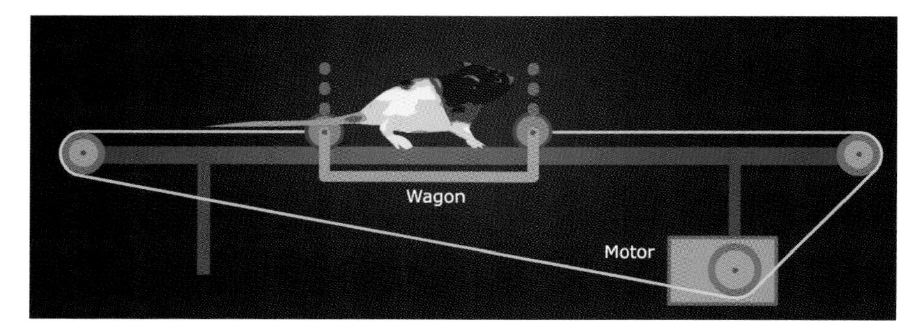

Figure 5. Emilio Kropff struggled for a long time to design an experiment that would allow him to control the rat's speed. In the end, he came up with the perfect solution: the Flintstone car. Illustration: Arnfinn Sorensen, Forskning.no

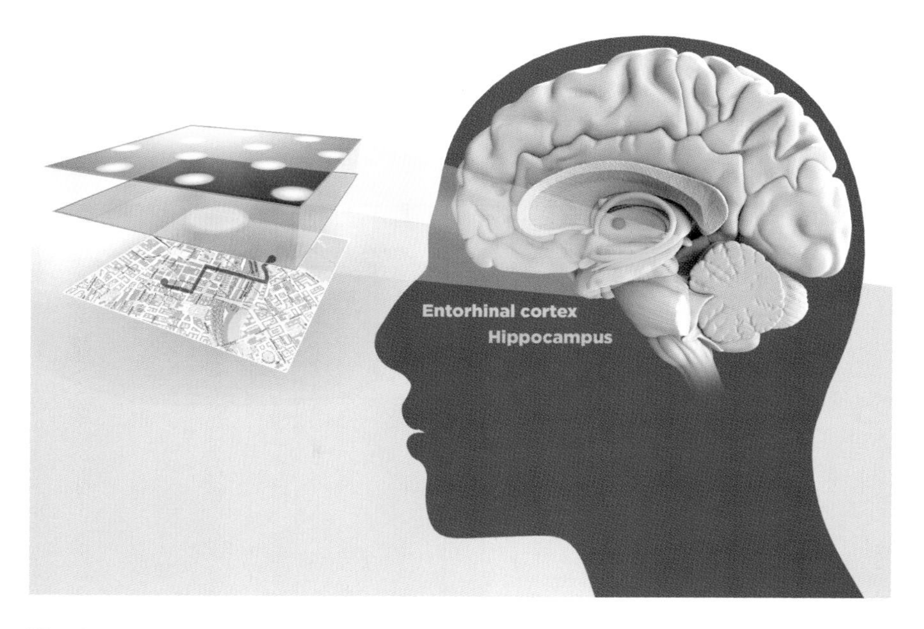

The place cells in the hippocampus represent memories of places, whereas the grid cells in the entorhinal cortex give us a coordinate system that tells us where we are. Copyright: The Nobel Committee for Physiology or Medicine. Illustrator: Mattias Karlén.

Nobel chaos at the Kavli Institute on October 6, 2014. Photo: Ned Alley/NTB Scanpix.

A victory toast with close colleagues at the Kavli Institute. From left: Clifford Kentros, Menno Witter, May-Britt Moser, and dean Stig Slørdahl. Photo: Nancy Bazilchuck, NTNU.

The day after news of the Nobel Prize breaks, Edvard Moser lands at Værnes and is greeted at the airport by May-Britt, daughters Isabel and Ailin, colleagues, and journalists. Photo: Ned Alley/ NTB Scanpix.

John O'Keefe, May-Britt Moser, and Edvard Moser at the Nobel Prize awards ceremony in Stockholm in December 2014. Credit: ©Nobel Media. Photo: Niklas Elmehed.

Norway's crown prince and princess visit the Kavli Institute at NTNU in 2015. Receiving VIP guests has become part of everyday life for the Nobel Prize-winning researchers. Photo: Thor Nielsen, NTNU.

CHAPTER 11

A Pattern Emerges

Some weeks after Fyhn's article was published, California-based hippocampus researcher Bill Skaggs came across it. Skaggs immediately noticed one of the figures, Figure 2B, which showed where in the environment the cells were active. The figure struck him as extremely unusual. He had seen tens of thousands of figures from hippocampal cells over the course of his career, but he had never seen anything like this.

Skaggs took a screen shot of the figure and opened the image in GIMP photo-editing software, on which he worked as a developer in his spare time. Over the next few hours, he sat at his PC, absorbed in the figure and manipulating it in different ways. He gradually became more and more fired up. Wasn't that a pattern he was seeing? He overlaid it with a pattern of densely packed triangles to see how the figure fit into it. The fields fit almost perfectly into the corners of the triangles. Then he tried overlaying a hexagonal pattern – densely packed with six-sided figures. And that nearly blew his mind. Skaggs had done a lot of work on theoretical models of place cell activity in the hippocampus. This figure, if he was right about what he was seeing, matched certain theoretical models of place activity[45] and would revolutionize the entire field of hippocampus research.

"I wasn't sure whether the Moser group realized what they had and how important it was. At the same time, I was afraid other people had seen the same thing as me," Bill Skaggs says.[46]

Skaggs was afraid other researchers would try to pre-empt the Moser group with an article describing the pattern. That could lead to controversy about who had been the first to discover it. At first, he wasn't sure what to do, but after discussing it with some colleagues, he concluded that the most sensible thing was to send an e-mail to the research group in Norway.

He sent a figure from the *Science* article as an attachment to his e-mail, but with lines sketched in that ran in and out of the figure. These lines showed that the fields fit into a pattern of repeating equilateral triangles.

"We saw it at once then. Wow, that's what it is, of course," Marianne Fyhn says.[47]

This was the pattern they had been searching for!

In the e-mail, Skaggs wrote: "This is of course very remarkable. I think it is not an understatement to say that this is the most remarkable piece of data in the history of hippocampal research (with the possible exception of HM's amnesia) and immediately obsoletes all existing theory about how the hippocampus represents space."[48]

To confirm that there really was a pattern, and that it was as Skaggs proposed, they immediately began to test rats in a larger box. They had another, circular box in the lab, earmarked for another experiment, which had a diameter of 2 meters. Now they rigged it up and set a rat to work exploring the box.

They saw it in the very first experiment: The recordings in the 1 × 1 meter box had given them only a fraction of the picture, but now it was as if they had zoomed out and had a better overview. There really was a pattern. The cells' activity fields were neatly distributed across the box at regular intervals (Figure 3).

"Next we measured this extreme regularity. The angles are 60 degrees, they're equilateral triangles. Once we understood the scope of it, it was absolutely wild," Marianne Fyhn says.

Now everything happened at a furious pace. Edvard and May-Britt arranged to meet Skaggs at the annual neurological gathering of the Society of Neuroscience, due to take place in San Diego at the end of October 2004. At the same time, Marianne Fyhn and her husband Torkel Hafting decided to alter the poster they would take along to the conference. Hafting had been brought into the project as Fyhn's replacement when she was on maternity leave with the child they'd had in February. Their original plan had been to show that the cells retained their properties even when they were tested in the dark. Now, in addition, they would show the research community the incredibly regular pattern that had appeared during the recordings in the large box.

Some weeks later, Edvard and May-Britt were sitting with Bill Skaggs in the breakfast room of a conference hotel in San Diego. The Norwegians were jet-lagged and had piled their plates with potatoes, omelets, and Mexican food from the breakfast buffet, as well as coffee and fruit. The topic of the meeting was the amazing hexagonal pattern they had discovered. In mathematics, a pattern like this is referred to as both triangular and hexagonal, since six triangles combine to form a six-sided figure – a hexagon.

These kinds of hexagonal patterns appear throughout nature. Bees produce hexagonal patters in beeswax, snow crystals are six-sided, volcanic structures like the Giant's Causeway in Northern Ireland form six-sided pillars, and millions of chemical molecules are composed of six-sided carbon rings. Densely packed circles also form a hexagonal pattern. If, for example, you were to arrange coins so that they were touching on a sheet of white paper, and draw a dot on the sheet wherever the coins touched each other, the dots would form a hexagon. The six-sided patterns that occur in nature probably arise when cells are packed together as efficiently as possible.

While they ate, they discussed theoretical models that might explain how the place cells' activity in the hippocampus came about.[49] Skaggs explained to them how the models could lead to a hexagonal pattern.

"It's one of the most important meetings we have had in our career. The fact that the pattern was hexagonal was one thing; but its hexagonal form only became interesting to us once we realized how it could arise," Edvard Moser says.

May-Britt and Edvard were enthusiastic and wanted Skaggs to work with them, but he was busy with other projects and said it wasn't necessary either. When they returned to Trondheim, they had a list of experiments and analyses they needed to do to confirm that the pattern truly was hexagonal and see how it responded to inputs from the environment. In addition to the advice from Skaggs, they had also received useful input from numerous researchers in the field. The fact that they dared make their research results public at such an early stage was one of the keys to their success.

"These Society of Neuroscience meetings were extremely useful for us because we chose to go against the current. The trend has been that people are afraid to make data public when it isn't ready to be published. We came out with data whose meaning puzzled us. That allowed us to have fantastic discussions with the international academic community, then travel back home and produce some great work," May-Britt Moser says.

Over the following months, everybody on the team worked flat out. Marianne and Torkel did the experiments, and when they were away, May-Britt stepped in. One of the experiments involved rotating the box 90 degrees to see if the firing pattern changed. The cells responded by rotating the pattern by the same number of degrees, showing that the pattern was anchored in physical landmarks in the animal's surroundings. When they tested the rats in the dark, the same pattern emerged, even though the rats couldn't see where they were. This was confirmation that the animal was keeping a kind of internal overview of its location.

Alongside the lab experiments, Sturla Molden was working to produce software that could confirm the hexagonal pattern. To be absolutely certain that they were dealing with such a pattern, they had to use something known as spatial auto-correlation – a mathematical tool that finds repeating patterns that may be hidden by noise.[50]

May-Britt was impatient and constantly asked how soon he'd be finished. One Saturday in late autumn, Molden had finally reached the finish line when May-Britt stuck her head around his office door. Excited, she hurried to her own office and opened the program. She pulled up one of the recordings they'd made and ran it through auto-correlation. The picture that appeared on the screen resembled a round rug with a pattern of yellow dots on a turquoise background. The space between all the dots was the same, but every other row was slightly displaced, so that the dots ended up in between rather than directly above or below the dots in the neighboring rows. It was incredibly regular (Figure 4b).

"This can't be true! Sturla you must have built in a bug," was May-Britt's first reaction.

May-Britt had to check how the program dealt with recordings from other brain cells too, to ensure that not everything that was run through the program would produce the same pattern. She pulled up recordings from hippocampal cells and ran them through Molden's program. None of them produced anything like the regular pattern of the cells in the entorhinal cortex.

"The idea of this kind of pattern, I don't think anybody in the world could have imagined this. They were so regular that it was easy to think that there must be something wrong with the equipment or something," Edvard Moser says.

Molden went through each of the 100 cells for which they had recordings. All of them had the same triangular or hexagonal activity patterns. There was no doubt that the pattern was real.

Soon they felt confident that they had enough data to publish their finding. Around Christmas 2004, they had a visit from Bruce McNaughton and Carol Barnes, who were both working as guest researchers at the Institute. Marianne Fyhn presented the results to their guests, and the team realized from the visitors' reaction that they were onto something really big.

In February 2005, they sent the article to the leading journal, *Nature*. They considered several different names for the newly discovered cells before ultimately settling on grid cells, since the cells divided the world up into a grid pattern. In May 2005, the reviewers were satisfied, and the article was published.[51]

"Researchers in Trondheim have found the sense of place in the brain," anchors reported on NRK's news program on 1 August 2005.[52] A report from the Moser lab in Trondheim showed a rat with electrodes on its head running over a long belt in the laboratory. With the rat in the background, Edvard Moser told the reporter they had found brain cells that help us orient ourselves in space. "A tiny center in the rat's brain continuously computes its location. And since rats are pretty similar to us, this means people also have centers like this," explained the reporter.

In the report, Edvard explained how the brain subconsciously calculates distances and angles, and combines this information with data about landmarks in the environment, like a tree or a mountain. In this way, the brain forms an internal map of the environment.

Although they had now found and described the pattern of the map, they were a long way from understanding it. They had found a yardstick for the brain's inner map – a yardstick researchers had been seeking for many decades. Researchers worldwide were flabbergasted by the incredible finding, but May-Britt and Edvard Moser didn't have much time to celebrate. They were the first people in the world to see these cells, and now there were thousands of new questions to answer. One of them was whether there were other types of navigation cells that had yet to be discovered.

CHAPTER 12

Rats in Jackets – The Hunt for Navigating Cells

130 years after Charles Darwin proposed the existence of a specialized brain area that allows humans and animals to orient themselves in space, May-Britt and Edvard Moser had found this area in rats' brains. And since both rats and humans are mammals, and have the same fundamental brain structure, there was every reason to believe that humans also had such an area. An area deep inside both hemispheres of the brain and far removed from the outer senses was tightly packed with cells capable of measuring the world. But were the grid and place cells alone enough to constitute an animal's internal positioning system?

There was substantial evidence that other cells were also involved. When researchers modeled how the place cells in the hippocampus functioned, they found that information from the grid cells in the entorhinal cortex was insufficient.

In 2004 Francesca Sargolini, an Italian researcher, came to the Moser lab as a postdoc. Along with Edvard and May-Britt, she started to research the entorhinal cortex area more closely. The entorhinal cortex consists of six layers of cells and fibers, and the grid cells they had published about in 2005 were located in the uppermost layer. The question was whether there were also grid cells in the lower layers.

This was the starting point for Sargolini's project. When she started to record in the deeper layers of the entorhinal cortex, she noticed that some of the cells appeared to become active as soon as the rat turned its head in a particular direction. Her suspicions were confirmed when she put together the images she had taken with a camera and the recordings from the brain cells: These had to be head direction cells of the kind James Ranck had first discovered in 1984 – only in a different but nearby area of the brain.

She showed the results to Edvard, but he wasn't very interested at first. He told her that head direction cells like this were probably scattered all over the place. Sargolini went back to the lab and did some new experiments. The head direction cells continued to appear in the recordings. There weren't just one or two of them: It looked as if there was a large concentration of these cells in the deeper layers of the entorhinal cortex. She went back to Edvard and showed him the new results. "Do you still think it's just a few?" she asked. He leafed through the printouts of the recordings. "Amazing," he said when he saw what she had found.

There were more surprises in store. It turned out that many of the cells found by Sargolini had dual properties. They were simultaneously head direction cells *and* grid cells. In other words, they were capable of dividing their surroundings up into a hexagonal pattern and simultaneously knowing what direction the animal's head was pointing in. In one of the deep layers, as many as nine out of ten grid cells were also head direction cells. They called these conjugated cells.[53] But what were the implications of the fact that some cells had both properties simultaneously?

"We didn't expect to find these cells at all. But once we had, it was clear to us that there was an advantage to having both types of information in the same system," Francesca Sargolini says.[54]

Some theoretical models, including one developed by Bruce McNaughton, had also predicted that such cells might exist. In order for an animal to be able to move about in a logical way, it must know which direction it is going in and its current position – and so it is advantageous for this information to be integrated into one and the same cell.

Other models suggested that there could be even more cell types in the brain's positioning system.

BORDER CELLS

The researchers who recorded cells in the entorhinal cortex became accustomed to seeing a lot of odd cell types. The grid cells accounted for only around half the cells in the area. That was one of the reasons why Edvard wasn't initially surprised that Sargolini had found head direction cells. In order to complete the projects they were working on, the researchers couldn't let their work be held up by cells other than the ones they were focused on. They simply had to note other observations and wait until they had enough data to make sense of them.

One of the observations Edvard and May-Britt had noted in many of the researchers' data material was a type of cell that appeared to react to borders in the environment. Marianne Fyhn, Torkel Hafting, and Francesca Sargolini had all seen this type of cell in their data. Edvard and May-Britt found these cells particularly interesting because they were aware of a theoretical model developed by John O'Keefe and Neil Burgess that suggested the existence of a cell type that reacted to borders. The model built on experiments showing that if a box was expanded in one direction, the field of the place cells stretched, as if the field were somehow "hanging on to" the walls. One possible explanation for it could be the existence of cells that indicated to the animal how far away it was from the different walls and which direction it was facing in relation to each wall. O'Keefe and Burgess called these theoretical cells boundary vector cells, but they had not yet been found in experiments.[55]

Another person looking at these cell types was Trygve Solstad, a levelheaded man from Trønderlag in northern Norway. Solstad, whose background was in mathematics and psychology, had joined the Moser group as a master's student in 2003 and gone on to do a PhD. While recording from grid cells, he came across a cell type now and then that only appeared to be active along walls. He found these cells interesting because they were so clear and consistent.

Early in 2007, Solstad presented his observations of these border cells at one of the weekly lab meetings. Immediately, May-Britt and Edvard set him to work looking for more such cells and mapping them, along with fellow researchers Charlotte Boccara and Emilio Kropff.

At the outset, it was simply a matter of devising experiments that would confirm whether these really were cells that reacted only to borders. Working with a dedicated team of technicians – veteran Kyrre Haugen and electrical engineers Klaus Jenssen and Endre Kråkvik – Solstad spent a long time designing the perfect box for the project. It had to be extendible in both directions, but in such a way that the rat would think it was still in the same box. No olfactory or visual markers could be moved. They also had to ensure that the rat could not detect the sliding mechanism with its whiskers. And finally, the box had to be easy to clean.

"When you see research on TV, it all looks so professional and simple, and people have all kinds of instruments and measuring equipment to hand. But in real life, those things aren't there before you start out on your own study; they have to be made," Trygve Solstad says.[56]

Just as they were ready to start on the experiments after many months of preparation and animal training, the lab happened to have a visit from Neil Burgess of London – who, along with John O'Keefe, had proposed the boundary vector cells theory. Although he was working in a competing lab group, it was natural for Edvard and May-Britt to invite him into the laboratory to observe the border cell experiments.

"There was never any secrecy. May-Britt and Edvard always practice openness. Never in my experience did we hide results from one another or refuse to let researchers into the lab," Solstad says.

Solstad and Boccara had set up experimental boxes in different rooms of the lab in order to observe how the cells they were recording behaved in different spaces. In one room, Room 3, the cell was active in a narrow strip that ran parallel to the entire western border of the square box. Solstad asked Neil Burgess whether he had any suggestions for an experiment they could try out while he was there. Burgess was keen to test an experiment Solstad had planned: What would happen when they placed a new wall into the box, which ran parallel to the wall the cell reacted to? If the cell they had identified as a potential border cell really was reacting to borders, it should start to fire on the same side of the new wall as it did on the fixed wall. This was an absolutely conclusive test that the cell would have to pass if it genuinely was a border cell.

Solstad lifted the rat, numbered 12018, out of the box and added a new wall. He also scattered rewards across the floor of the box to encourage the rat to explore the area as much as possible. Once he had checked that they were still picking up good signals from the cell, Solstad gently placed the rat back in the box.

As the rat started to sniff around, Solstad and Burgess monitored what was happening in the cell. It crackled and popped every time the rat neared the western wall, just as it had before. When it explored the space between the new and the old wall, however, the cell fell silent. And when it approached the new wall from the west, the cell remained silent. But as soon as the rat rounded the end of the wall and found itself east of the new wall, the cell abruptly woke up again and fired a small electrical charge. The same thing happened every time the animal approached the new wall from the east. When they removed the wall again and let the rat explore the box one more time in a control experiment, they saw that the activity vanished again. The cell really was reacting to the wall.

"It was really cool. It was the first time we knew we were dealing with a genuine phenomenon. Now we were confident there were cells that are interested in walls and that the directional placement of the wall does matter," says Solstad.

Over the weeks and months that followed, the border cells passed one test after another. The field of the border cells stretched when the box was expanded in the direction the border cell was interested in, but didn't change if the box was expanded in a different direction. If the rat was moved to another space, the border cells continued to fire along walls in the new space; these weren't grid or place cells that had randomly fired along a wall. When Solstad removed all the walls around the box, simply leaving a free fall along the border, the cells still fired along the borders of the box. These cells genuinely deserved the name border cells.

After a while, the experiments became routine. The researchers sat in the semi-darkness scattering chocolate cookie crumbs for the rats. But it paid to stay alert and encourage the rat to cover the entire surface area of the box. That is why Solstad was determined that every experiment had to be absolutely impeccable. Along the way, he regularly presented his results to May-Britt and Edvard, discussing with them what experiments needed to be done and whether the data were convincing enough. Just a few months after launch, they were ready to publish, and in November 2008, the article about these cells appeared in *Science*.[57] The Moser lab was now responsible for discovering two whole positioning cells.

Around the time they discovered the border cells, the center's standing in the research community enjoyed a major boost. Recently, the Norwegian billionaire and philanthropist Fred Kavli had visited Norwegian universities on the lookout for exceptional research communities that might qualify to become so-called Kavli Centers, thereby receiving lifelong research support. Kavli had set up 14 such centers worldwide within the three academic disciplines of astrophysics, nanoscience, and neuroscience, but none of these centers were in the Nordic region. At NTNU Trondheim, Kavli and the then president of the Kavli Foundation, David Auston, were brought to Edvard and May-Britt's laboratory at the Medical-Technical Research Center. Kavli and Auston were impressed by the pair's achievements, especially when Edvard and May-Brit showed them the figures for an as-yet unpublished grid cell article.

After the visit, the department was invited to apply to become a Kavli Center. The application was approved, but there was one hitch: Fred Kavli had made the gift conditional on receipt of equivalent financial support from the Norwegian side over a number of years. Nobody could offer this kind of guarantee. Norway operated only with one-year budgets, and it was difficult for the Ministry of Education and Research to promise them multi-year funding. Even so, the then chancellor, Torbjørn Digernes, managed it. On August 14, 2007, the Center for Memory Research was named the world's 15th Kavli Center and the first such center in the Nordic region. In addition to serving as a seal of quality, this also secured research resources that would enable the center to continue its scientific research for a long time to come.

SPEED CELLS

From the time when place cells were discovered in 1971, it had taken more than 30 years to find another cell type that might form part of a cognitive map. And now suddenly, they had discovered two entirely new types of positioning cells in just four years.

"There was a long run of successes. It all happened so quickly that we almost had trouble keeping track of it ourselves. Over three or four years, we discovered an enormous amount. One discovery followed another, and we had the field to ourselves. Nobody else was working on the same topic, and we had no competition. It was like unwrapping a Christmas present – everything was revealed," Edvard Moser says.

Rats, and probably all mammals, appeared to have a kind of GPS system built into their brains. The GPS had grid cells that divided the surroundings up into triangular fields and cells that informed the animal which direction it was going in and where the boundaries of the environment lay. But there was still one missing factor: speed.

Several research groups that studied the hippocampus believed that they had seen indications of cells that reacted to speed, but nobody had managed to produce any convincing data. It was therefore assumed that the brain had no way to record speed. But how could the brain and the GPS get by without a speedometer?

One of the people interested in this question was Emilio Kropff of Argentina. He was a theoretical physicist who had spent much of his career building computer models of the brain. In 2003 he was a PhD student in Italy under Alessandro Treves, a physicist who was collaborating with the Moser group. In 2004, Treves returned from Trondheim and told Kropff about the as-yet unpublished discovery of the grid cells with the hexagonal pattern. Kropff could hardly believe his ears.

"To any scientist, it's just mind-blowing to hear that there are neurons deep inside the brain with such a precise perception not only of where the animal is but also of geometry. That a single neuron can link different positions so that this neuron will answer in precise locations forming a hexagonal pattern. It was beyond our wildest dreams," Kropff says.[58]

After that, Kropff couldn't get the grid cells out of his head. How could a pattern like this come about? Although he was busy with a PhD project about something entirely different, he started to mull over the problem with his supervisor, Treves. It started off as an amusing mystery, but the question gradually occupied more and more of his time. Thoughts about the pattern woke him in the middle of the night.

In 2008, he co-published an article with Treves in which they described a new theory of how the pattern might arise. Since there was, at that point, nothing to indicate that speed was being directly coded in the brain, they suggested that the solution was to record time. They pictured the grid cells as being like miniature pacemakers that fired with a rapid pulse, just like the cells in the heart. The animal's sense of space could therefore come from matching sensory cues with the rhythm of its own inner chronometer. It would be easy to test Kropff and Treves' theory: If it was correct, the distance between the points in the pattern would differ with differences in the rat's speed. Even before the article appeared, Kropff had written to May-Britt and Edvard Moser, and asked to come and test his theory with them.

"At that time, it was still possible to get into their lab even if, like me, you hadn't published any super-important articles. It got so competitive within a few years that it became very hard. It was a great moment, the day I received the e-mail from May-Britt saying they were interested in me. It was a day I'll never forget," Kropff says.

In 2008, Kropff moved to Trondheim in the hopes of proving his theory. Nobody knew if he would be any good at operating on and training rats, since he had no experience working with lab animals. But thanks to close supervision, he gradually became skilled at it.

In May-Britt and Edvard Moser's lab, he had time to think through how best to set up his experiments. This was something the Mosers encouraged their researchers to do. They could also spend more time than other research groups on each project thanks to their long-term research funding.

"The Moser lab is one of the only places in the world where you can actually have long stories. In general, in scientific labs, you're desperate to get data so you have to hurry and find the experiment you want to do very fast. It was a great experience not to be in such a hurry to get data," Kropff says.

In order to test his theory, it was crucial for Kropff to be able to control the rats' speed. The question was how.

RAT JACKETS

The simplest thing would be to place the rats in a remotely controlled cart that moved at a given speed and see how the rat brain reacted. But other research groups had already tried this without success. If Kropff was going to test his theory, the animal would need to move under its own steam.

He discussed many different ideas with both Edvard and May-Britt, as well as other colleagues. At first, he tried scattering marbles across the floor of the box so that the rat would constantly have to zigzag around to avoid them. The idea was that the rat would move more slowly than if it were running around in an empty box.

But in practice, the effect was the complete opposite. The rats found the marbles exciting and ran around and played a lot more than usual. It also turned out to be incredibly time-consuming to clean the feces off the marbles after every experiment – which was important both for reasons of hygiene and for preventing scent trails from being left anywhere in the environment.

After shelving the marbles, he tried to slow the rats by weighing them down with weights. On John O'Keefe's advice, he tried, among other things, to tape a small weight to their tails. But after trying it a few times, he discovered that the tape made the skin on their tails sore, so he had to discard this method too.

His next idea was to dress the rats in little jackets. If he sewed pockets into the jackets and then placed tiny lead weights in them, that might slow the rats' pace. He struck on the idea that socks could be a good starting point, so he got hold of some baby socks, cut off the soles, and sewed some pockets onto them. But he quickly discovered that making clothes for super-supple animals that can wriggle their way through the narrowest openings was hopeless. It was almost impossible to make a jacket that they couldn't slip out of. He also realized that the jacket couldn't cover the rat's underbelly. As rats run, their underbellies brush against the ground beneath them; when the jackets covered their underbellies, the rats' balance deteriorated. Kropff tried to alter the design, using tops that left the underbelly bare, but that didn't work either.

Despite these failures, Kropff remained convinced that slowing the rats down with little weights was the way to go. Since it was so difficult to attach weights to the animal itself, he came up with another idea: a pulley system set up so that the rat pulled a weight when it ran in one direction on a treadmill, and was then pulled by the weight on the way back. But this too failed.

Eventually he had to come to terms with the fact that none of his ideas were working.

THE FLINTSTONE CAR

Despite his frustration, Kropff wasn't ready to give up. He began to reconsider his initial idea – using a wheeled vehicle. What if he made a kind of wheelchair, placing the rat's hind legs in the chair and leaving its front legs to move freely? Kropff decided that would be his next experiment.

At this point, it was Kropff's turn to present the latest updates on his project at the Institute's lab meeting. Armed with all his failed experiments and this new idea,

Kropff went to the meeting and laid out his problems. One of the researchers there was Jonathan Couey from Menno Witter's research group. After listening to Kropff's presentation, he spoke up: "Why not build a car without a floor?"

Kropff let the idea sink in and realized that it was the perfect solution to his problem.

Some months later, all eyes were on a four-meter-long treadmill. Emilio Kropff had three world-famous scientists in his audience at the laboratory: The Nobel prizewinner Erik Kandel, who was still vigorous and alert despite being well over 80, John O'Keefe, and Richard Morris. Kropff lowered a lab rat into a frame fastened to the treadmill and started the motor. The frame began to drive across the treadmill and the rat kept up with it, running at the same speed inside the "car." It looked pretty much like Fred Flintstone "driving" around in his car, propelled by his own feet. The sight was hilarious, even for the blasé audience. They loved it and were impressed by the simple genius of the experiment (Figure 5).

"The car experiment just worked so nicely. Typically, it takes up to half a year to train an animal in a task. That's the tough thing about working on behavior. It isn't that the rats don't want to do what you want them to do, it's just that they don't understand it. But with the car, it was instantaneous. The rats understood what they had to do," Kropff says.

Kropff let the rats run half the length of the treadmill at a rapid speed and the rest of it at a slower speed, with an abrupt change of speed in the middle. As the rats ran, electrodes registered what was happening in the entorhinal cortex. In order to exclude place information from the experiments, he let the rats run the same stretch along the treadmill at a constant speed chosen by a computer – either 7, 14, 21, or 28 cm per second.

When Kropff went through the results, he saw no signs that the distance between the grid cells had changed as it should have according to his theory. Instead, he found that some of the cells appeared to adjust their firing rate in line with the speed. Ten different rats performed the experiments, and in all of them he found that a little more than a tenth of the cells reacted in this way. The faster the rat ran, the faster the cells fired. Slowly but surely, he had to acknowledge that his hypothesis was wrong. These were, of course, speed cells. The brain *did* have a speedometer.

"I was not expecting that at all. It was quite frustrating for me. When we found speed cells I had to change everything that I thought about grid cells. The Mosers weren't as committed to my theory, of course, but they were always very open to all kinds of theories, especially if the theories involved clear predictions that we could test," Kropff says.

May-Britt Moser confirms this.

"It has always been important for us not to wed ourselves to any one theory. It's fatal to do that as a researcher because then you may find yourself looking extra hard for confirmation of what you believe and overlook what contradicts the theory. But we don't throw the baby out with the bathwater either and give up on a theory straight away. There's always a balance. But if the data are good, the world goes on, we've learned something new, and that gives us enormous satisfaction."

Kropff also tested whether the speed cells behaved in the same way when the rats weren't using the Flintstone car. When he recorded the speed of the rats as they were moving freely in the box, the same thing happened. The speed cells fired faster when the rats were running than when they were wandering around slowly.[59]

Only after Kropff had come to terms with having torpedoed his own theory was he able to take pleasure in having found the speed cells.

"Before, people had seen some neurons fire more in some situations, but it was hard to convince other people of it. But I think we put together enough evidence to convince everybody that this was about speed."

With the discovery of the speed cells, they had mapped the final missing factor in the cognitive map. Kropff thinks it's only a matter of time before speed cells are discovered in humans too.

The hippocampus and nearby areas are key to both our positioning and memory systems, even though these systems are distinct. If the hippocampus is damaged, people lose both their spatial perception and much of themselves. Henry Molaison, H.M., taught us that.

May-Britt and Edvard's desire to understand the mechanisms underlying memory and learning had been their motivation for studying the brain. After mapping the cells that constituted spatial perception during the 2000s, they wanted to understand how spatial perception and memory were connected.

Why are spatial perception and the memory machine located in the same part of the brain? What is the mysterious connection between places and memories?

CHAPTER 13

The Memory Palace

To Greek and Roman thinkers living 2,000 years ago, it was clear that there was a connection between place and memory, and they used this connection to develop memory techniques. In *De Oratore*, a work that dates back to 55 BCE, the Roman orator Cicero advised speechmakers to choose a place they could conjure up in their thoughts and where they could mentally lay out different objects that would later help them remember what they wanted to say.[60] The method went by the name of "Loci" or "memory palace," and is used to this day by the world's foremost masters of memory to remember long series of numbers or cards.

PLACE CELLS OR MEMORY CELLS?

In the decade after John O'Keefe discovered place cells in the hippocampus, it became clear that these cells weren't mere place references for the rat's location at a given point in time. O'Keefe found that place cells could also represent where the rat *thought* it was.[61] Soon, other researchers showed that place cells encoded information about where the rat had been or where it considered going. Both in a waking state and while the rat was sleeping after an experiment, it was possible to see the place cells "switching themselves on" again, as if the rat were replaying memories of where it had been.

Another indication that place cells encoded memories of events came from a series of interesting experiments carried out by American researchers Robert U. Muller and John Kubie. Like James Ranck, the researcher who discovered head direction cells in the eighties, the pair worked at State University of New York. Muller and Kubie were interested in how different places were represented in the hippocampus.[62] How would place cells react when a rat moved from one environment to another? They envisaged one of two scenarios: Either the place fields would be entirely redistributed in a new space, or they would retain the same placement relative to one another as in a previous space. For example, place cell A would always have a place field to the left of place cell B.

To determine the correct model, Kubie and Muller trained the rats in four different environments: a small and a large round box, and a small and a large rectangular box. Then they recorded the place cells while the rats spent 15 minutes in each box in succession. The results showed that it was entirely random whether a place cell was active in a new environment after being active in the preceding one. One of the most surprising findings was that place cells that had been active and clear in one box could be silent in the next. At first, Kubie and Muller wondered whether something had gone wrong with the equipment and they had lost the recording of this place cell. But when they placed a rat back in the first space, the same place cells fired again. These cells simply hadn't been interested in the new space. The two researchers called this total reshuffle of place fields "remapping." It was as if the hippocampus made a unique map for every environment and kept a catalog of all the places the animal had been. Later work also showed that smaller differences between environments – for example a change in the color of a sheet of paper hanging on one of the walls in the box – could also cause the animal to create a new map.

Edvard and May-Britt were interested in exploring the remapping of place fields further. They wanted to understand what triggered it. Figuring out what happened when the hippocampus switched between two different representations of place could be the key to understanding the hippocampus's mechanisms for memory storage.

Edvard and May-Britt enlisted the help of two other researcher couples, Jill and Stefan Leutgeb, and Carol Barnes and Bruce McNaughton. Together they worked out a series of experiments that they hoped might reveal what was happening.

Jill and Stefan Leutgeb set up two separate experiments. In one, they allowed rats to explore boxes that differed from each other in color and shape (black versus white walls; round versus square). The boxes were always in the same place in the same laboratory room. These could be seen as different events in the same place. In the second experiment, they let the rats explore exactly the same box, but in two different laboratory rooms. This was meant to mimic the same event happening in two different places.

The rats had arrays of electrodes implanted in their hippocampi, so that the researchers could see how the cells reacted during the experiments. When Jill and Stefan Leutgeb analyzed the results, they found a clear difference between the two scenarios. When two identical boxes were placed in different laboratories, the rats stored different "maps" for the two boxes, with new and entirely independent sets of active neurons in CA3 – a kind of total reshuffle of place fields, which the researchers called global remapping. But when the rats explored two different boxes located in the same place, something different happened. In that case, the same cells were active, but their firing rate changed. It was as if the cells were saying that something was different even though the map was still the same. This meant that the remapping wasn't just one process; it was two.[63]

It seemed as if the brain created unique maps of all the different places the animal had been and switched from one map to another when the external surroundings changed. If there really was a map for every single memory, this reshuffling or remapping process had to be extremely important, May-Britt and Edvard Moser thought.[64]

In their next experiments, they would discover just how quickly the brain could switch between different maps.

TELEPORTING

Many people have had the experience of waking up in the middle of the night in a hotel room or guest room and not knowing where they are. For a few seconds, it feels as if the brain is working in top gear to find its way to something in your memory that matches the walls and furniture you can make out in the dark of the night. In 2006, May-Britt and Edvard collaborated with the Czech researcher Karel Ježek on an experiment to recreate this kind of spatial confusion in the laboratory. According to a hypothesis proposed by the neuroscientist John Lisman, an "alarm signal" would go off in the hippocampus if sensory information from the cerebral cortex didn't correspond to the stored memories the animal had of a particular place.

In order to find this alarm signal, Ježek tried to surprise the rats by switching the experimental boxes in different ways.[65] First, he introduced a round plastic container into the experimental box. He thought this would be roughly equivalent to a new sofa suddenly appearing in a person's living room from one day to the next. But the rats didn't seem bothered by the container, and nothing special happened in their hippocampal neurons either. So he made the box more eye-catching, with black and white stripes. But that didn't provoke any reaction either. Since it would apparently take more to surprise the rat, he introduced a battery-driven toy cat, a bit smaller than a rat. It had a butterfly on its head, could move, and even meowed. Surely the rat would react to that! And indeed it did – only not in the way Ježek had hoped. When the cat started to move and meow, the rat was at first visibly scared but then started to examine it. The rat's fear system appeared to be on high alert but Ježek had no instruments to record that. And the hippocampus remained silent.

Ježek also suspected that the problem with his experiments was timing. He had no control over when the rat noticed the new object. That made it difficult to know when the feeling of surprise happened, if indeed it did. In order to gain better control over the timing of the surprise, he introduced new light designs. The great thing about using light was that he could combine different elements of the two spaces by flipping a light switch. He made one box with white light-emitting diodes that lit up the box through the transparent floor, and another box with green strip lighting along the edges. In addition, he made a third box that incorporated both light designs, enabling him to combine different elements of the two spaces.

One day when he was due to work on these light designs, he took the elevator up to the third floor where the laboratory was located, as usual. When the doors opened, he stood there confused for a fraction of a second. He could almost feel the physical sensation of his brain searching for the correct interpretation of what had happened before everything fell into place and he realized where he was. He had ended up on the fourth floor, the level above his lab.

This experience gave him an idea for the perfect experiment: Instead of combining different elements of the two boxes, he would simply go from one to the other with the flick of a light switch. The effect would be the same as the one he experienced when he came out on the wrong floor. In practice it would be like teleporting the rat from one space to another.

Shortly afterwards, he ran the experiment in the lab. He had spent a long time training the rats and familiarizing them with the two different light boxes before introducing the third box, which incorporated both light designs. He could tell that the rats perceived the box as two different places depending on which light design was switched on, as different place cells in the hippocampus fired in either case. Up until then, the lights had never been changed while the rat was in the box.

As usual, Ježek gently lowered the rat into the box with white diodes in the floor. The rat was left to run around and search for cookie crumbs for roughly a minute before Ježek flipped the switch for the green box. The rat froze, as if it were wondering what was happening. Then it continued to hunt for food. Only initially did it show any sign of registering the abrupt transformation in its surroundings. It soon stopped being surprised and simply continued to sniff around in the box in search of rewards.

Though the animal did not show any signs of noticing that it had been teleported to another space, Ježek could nonetheless see that the brain was responding in the same way each time. In just a tenth of a second, the rat switched to the map for the right box. But he saw no sign of any alarm signal.

After six rats had undergone a grand total of 169 teleportation experiments and the team had analyzed the results, they were able to conclude that the rats never mixed up the maps for the two boxes.[66] The brain could switch between the maps at lightning fast speed, and the maps always remained distinct.

This is probably what we experience when we wake at night, confused: These different representations are competing inside our brains.

"It may seem strange that the rats got used to being teleported to another space so rapidly, but we probably experience similar episodes ourselves every day," Karel Ježek thinks.

If you fall asleep on the bus on the way to work, you will likely wake up in a new place. And yet your brain recovers quickly and grasps where you are. In fact this happens every time an animal wakes up: It must rapidly process new information and retrieve the correct map; otherwise it won't survive.

The hippocampus appeared to be able to store an infinite number of place maps. The question was how it managed to do this. A project Edvard and May-Britt ran with a young husband-and-wife research team from Kristiansand helped reveal the role the grid cells in the entorhinal cortex played in enabling the brain to store all these representations.

FLEKKEN'S FOUR MAPS

It was a 2004 Swedish TV feature on May-Britt and Edvard Moser that first showed Hanne and Tor Stensola that it was possible to do research as a couple. Just four years later, they themselves were researchers in the Moser group.

The couple met and got together in Bergen in the early 2000s. Hanne had studied psychology while Tor was completing his civilian national service, working in the music department of the library. Both of them had wanderlust, so they decided to study abroad.

After completing their undergraduate degrees in neuroscience in Dunedin, New Zealand, they won scholarships to pursue a master's program at Oxford University. Early on in their master's degrees, they contacted Edvard and May-Britt Moser to ask whether they could come work with them.

In 2008 they moved to Trondheim. Their project involved investigating the size of the grid cell pattern. From the moment the pattern was discovered, researchers had noticed that the size of the grid varied depending on where in the entorhinal cortex they recorded. But nobody knew whether the size increase was gradual or happened stepwise. The challenge was to record from enough grid cells at a time to be able to draw conclusions about the way a large number of grid cells behaved simultaneously. They needed to greatly increase the number of grid cells they could record simultaneously – and they had an idea of how to do it.

The limitation was caused by the electrodes. In the entorhinal cortex, researchers used tetrodes – small electrode holders with space for 4 × 4 electrodes. In hippocampal research, it was common to use a larger holder, with space for 12 × 4 electrodes. One important reason why researchers had been unable to use this kind of electrode holder in the entorhinal cortex was that the holder needed to be slightly angled in order to be surgically implanted in this area, and nobody had managed to do this with the large holder.

Hanne and Tor went to the electrical engineer Endre Kråkvik, one of the people who had built the electrode holders, and presented him with their problem. "We need to have this at an angle of 11.2 degrees. Can you do it?" they asked. Endre responded by bending the tip of the holder with his thumb. "Like this?" Hanne Stensola took the holder with her and implanted it in a rat's brain. It worked right away. And with that, Hanne Stensola managed to do something nobody else in the world had yet achieved.

Whereas people had previously managed to make recordings from around twelve grid cells in each rat, the number of cells was now increased around tenfold. In one of the rats, they managed to record no fewer than 186 grid cells simultaneously.

May-Britt believes Hanne's musicality contributed to her success, enabling her to listen to the different cells and stop at exactly the right place. In addition, the surgery had to have been elegantly performed and the coordinates, perfect.

After a fairly short time, they saw that the grid cells produced at least two different-sized patterns when the rat was exploring a box. The smallest pattern "divided up" the floor of the box so that there were distances of around 35 cm between each place where the grid cells responded. Another group of grid cells produced a pattern for roughly every 50 cm the rat covered. They also saw indications of larger patterns, but the box they were using was too small to be sure that this was the case. To see whether they could find a third pattern size, they put the rat in a box whose sides were 1.5 meters long rather than 1 meter. This enabled them to see a third and fourth pattern. In the largest pattern, there was a distance of 172 cm between each point at which the grid cell reacted.

It looked as if the grid cells were continually dividing up the environment in at least four different ways. But could all this be interconnected – or did the different map sizes operate entirely independently of one another?

In fall 2009, Hanne and Tor discovered something peculiar. As they sat at the computer looking over the grid patterns they had recorded in one of the rats, they noticed that one of the patterns was slightly distorted. It was no longer perfectly hexagonal – it looked as if somebody had compressed it. At first, they didn't pay too much attention to it, because they knew that other people had seen similar deformations of the pattern, and that it was thought to be noise in the recorded data. But when they pulled up the recordings from more cells, they noticed that a different-sized grid pattern from the same recording session was also compressed, but elongated in a different direction.

This was odd. The deformed patterns could have nothing to do with the rat's general behavior or perception; otherwise, all of the patterns would be deformed in the same way.

"One of the greatest moments in our work was when we realized that the patterns might also be able to react differently to changes in the environment," says Hanne Stensola.[67]

Their hope was that if they switched the box, one or more of the grid patterns would change. This might also help them understand the principles underlying these changes.

They needed an experiment that could put the grid patterns to the test. Among other things they tried switching out the square box with a triangular box, to no avail. While they were doing this, May-Britt and Edvard were both off traveling, so they sent an e-mail to Edvard detailing their different attempts to change the environment and asked whether he could make any suggestions. He answered: "Can't you just move the wall?"

They realized at once that this was the simplest and best way to do it.

Although it was a Saturday afternoon, they immediately went back to the laboratory in Øya and moved the wall of the box. Hanne went to fetch their trusty rat Flekken ("Patch" in Norwegian), named for the large mark on his back. As always, he stood up on his hind legs when Hanne came into the animal pen, happy and eager as a dog setting its eyes on its master. He was well aware that she and Tor were "his" researchers. They had worked so much with him that they viewed him as a colleague. For his part, Flekken had come to trust them so much that he often lay down to sleep in their hands. Rats can be jumpy and nervous, hyperactive or listless. Flekken was a calm, good-natured rat, keen and motivated to work in the box. What's more, he had extremely good, clear grid cells that always gave them excellent recordings.

Hanne carefully lowered Flekken into the box, where they had now moved one of the walls 50 cm inward, making it rectangular instead of square. The rat set to work straight away, exploring the surface of the floor in the box and picking up small rewards that Hanne had scattered around for him. As he was doing this, she recorded the activity of his grid cells. Once she was finished, she and Tor pulled up the recordings on the computer screen.

The grid cells that produced the smallest pattern didn't seem affected by the fact that the box had changed. The pattern looked the same as in the square box except that it lacked the fields that now lay on the outside of the box. The grid patterns with larger gaps between the fields were a different matter entirely. In this case, there was a significant difference compared with the recordings from the big box. Whereas they had produced a perfect hexagonal pattern in the big box, the pattern in the rectangular box was compressed. They looked at one another and screamed with joy. Here was solid proof that the grid cell patterns came in different sizes and that the patterns were independent of one another.[68]

But what did the results from Flekken and the other rats actually tell them about the grid cells and their function? When researchers in the Moser group first found the grid cells, they all believed that the regular pattern could represent a yardstick that divided up the world. Now it was clear that the distance between the grid fields could change from space to space. Is it possible for the mammal brain to rely on such an erratic yardstick?

"Absolutely, I'm convinced of it. Our brain demands so much energy that I think anything it is capable of producing is used to the full. You can actually use an elastic band as an instrument of measurement, as long as you keep it stretched at a constant length in the environment you want to measure. So if the dispersal of the grid cells is different in two different spaces, that doesn't matter, as long as they remain constant in the space where you are," May-Britt Moser says.

It is enormously advantageous to have four different map sizes. If our brain is to store all the events a human being experiences in a lifetime, it needs an almost inexhaustible source of different place codes, or place labels. And maybe the grid cells are what give us these place labels. If each of the patterns can have many possible settings, the resultant combinations can generate an enormous number of different place codes. We could compare it with a bike lock that has four different wheels, each with ten possible settings from 0 to 9. A combination lock of this kind can be set in 10,000 different ways. If we imagine that each of the grid patterns can also be altered in an equivalent number of ways, this may give the brain the potential to assign every memory a unique place code. Perhaps spatial information was used as a kind of frame in which to store memories. This could explain why memories are often linked to places.

The connection between places and memories is probably ancient. Whereas some scientists think humans are alone in remembering events,[69] a number of psychologists and neuroscientists take the view that episodic memory may have a much longer evolutionary history, and may have evolved at several points in time.

Studies of magpies have given psychologists and neuroscientists probable reason to suspect that we are not alone in possessing episodic memory. Magpies have what scientists call *what-where-when* memory, which closely resembles our episodic memories.[70] English scientists discovered this by letting magpies hide small food balls in different colors – blue or red. The magpies were allowed to collect and hide as many balls as they wanted before being removed from the laboratory. Either that day or the following day, they were brought back into the room. If they came back on the same day, the blue balls were replaced with wooden balls of the same size and color, whereas if they came back the following day, the red balls were switched out. After a certain time, the magpies learned which food balls would be replaced with wooden balls and headed straight for the containers where they knew they would find food. They were not only remembering where they'd hidden the food, but also whether they had hidden it on the same day or the day before.

This experiment and similar studies of birds led neuroscientists Norbert Fortin and Timothy Allen to propose that episodic memory must have come about at a point in evolution before mammals parted ways with birds.[71] Even though this was millions of years ago, and our forefathers were a kind of lizard mammal called therapsids at that point, there were already species with traces of the brain structure that would later become our hippocampus. This suggests that the foundations of episodic memory may already have been laid by that point.

Edvard Moser is also inclined to believe that humans are not alone in having episodic memory. It appears that place and events are always linked and that this system has been extremely effective.

"I and many other people think that the development may have started with a system for finding one's way. All animals must be able to do this; it's absolutely critical to survival. And perhaps the entorhinal cortex and the hippocampus were important for this. But then a side-effect of the system has been that it can also be used to store memory," Edvard Moser says.

"If what we think is true – that the hippocampus can produce many maps, maybe hundreds of thousands of different maps based on the signals it receives from the entorhinal cortex – that makes it very useful for memory. Maybe this means that the expansion of the memory actually came from having this kind of system for spatial perception. We've always had memory naturally, even the simplest species. But episodic memory, the ability to remember events, may have expanded tremendously when the hippocampus and the entorhinal cortex evolved."

But behind the question of how memories are encoded in the brain lies another, more fundamental one: What is the point of memories? Why have we developed the capacity to store memories?

The Mosers' colleague and hippocampus expert, Menno Witter, doesn't believe memory in itself is the primary task of the hippocampus.

"Why do we care about memory? The reason we care is because we need memory to evaluate the outcome of our behavior. We need memory to predict the future. So I think that was what the hippocampus is doing, and why the hippocampus is so crucial for humans, and why it is still there. It is because we use it every single day to predict the outcome of our behavior," Menno Witter says.

Regardless of how places and memories fit together, some places and events are more firmly embedded in our memory than others. For example, winning the world's most prestigious science award, the Nobel Prize.

Chapter 14

"Ring Me At Once"

When May-Britt Moser climbed the four flights of stairs at the Norwegian Brain Centre on Monday October 6, 2014, hair loose and a woolen cardigan over her floral print dress, she had no idea how memorable this day would be. Many people had taken time off for the autumn break, and it seemed quieter than normal at the Kavli Institute. Edvard Moser was en route to a meeting at Munich's Max Planck Institute. May-Britt hurried through the door on the fourth floor as usual. On Monday mornings they had lab meetings, where they took turns to present results from the various projects they were running, and she was looking forward to hearing the latest news from the other researchers.

That same morning, Hege Tunstad, head of communications at the Kavli Institute, dropped by the office of the head of the Faculty of Medicine, Dean Stig Arild Slørdahl. Tunstad reminded Slørdahl that the winners of the Nobel Prize in Physiology or Medicine were due to be announced in Stockholm that day. "Should we be on standby?" she asked. "No. We can just relax this year," Slørdahl replied. May-Britt and Edvard might be potential Nobel candidates in the long run, but given the average age of Nobel prizewinners, the prospect was probably 20 years away. Besides, the prize had gone to neuroscience the previous year, and it was highly unlikely that the same field would be chosen again this year.

At around ten-thirty, when May-Britt and the others were at their lab meeting, absorbed in an interesting discussion about the research results presented that day, her phone rang.[72] She checked and saw it was an unknown foreign number. For a moment she considered not answering, but then she decided she'd better take the call and left the meeting room. The caller was a man, who spoke in Swedish and introduced himself as Göran Hansson, chairman of the committee for the Nobel Prize in Physiology or Medicine. May-Britt's first thought was that he wanted to ask her for a comment on somebody who had won the Nobel Prize.

She managed to get into her office before he continued. Instead, he told her that she and Edvard had won the Nobel Prize jointly with John O'Keefe, and that the news was strictly embargoed until the public announcement at eleven-thirty.

May-Britt couldn't believe her ears, even though he repeated the news several times. She asked him to send her an email so that she could see it in writing. Hansson said he would, then advised her to have a coffee, calm down a bit, and plan what she would say to all the journalists who would soon turn up. An email arrived immediately afterwards. She opened it and read: "Nice speaking with you a few minutes ago and congratulations on your Nobel Prize! We are truly delighted to be able to award it to you; this is a great day. I have tried to reach Edvard, but his cellphone was switched off, so I left a message telling him to ring me – or you. Let's hope he has a chance to call before the journalists arrive!"[73]

May-Britt burst into tears. She rang Edvard, but his cellphone was off. She rang their daughters. Isabel didn't pick up, so she sent an SMS that simply read: "Nobel OK. Come." She got hold of Ailin and told her to come. She didn't take the comment about journalists all that seriously: This is Trondheim – journalists don't just turn up here, she thought.

While all this was happening, Edvard was still on board the plane to Munich with his cellphone switched off, absorbed in work on an article he was writing.

At eleven o'clock, May-Britt went down to the second floor for a scheduled meeting with Stig Arild Slørdahl. "You need to sit down, Stig," May-Britt said. And then she read him the e-mail she'd received from Göran Hansson. He jumped out of his chair and congratulated her. It took a little while for them to pull themselves together and alert the chancellor, Gunnar Bovim, who was in Oslo that day. Slørdahl had a running joke with Bovim that he would retire if May-Britt and Edvard got the Nobel Prize, so the SMS he sent to the chancellor read: "I hereby announce that as of eleven-thirty I will be retiring on a solid pension from NTNU forever after... absolutely confidential until eleven-thirty." Two minutes later, he sent a selfie with May-Britt. Since Bovim didn't get the hint, Slørdahl realized he'd better ring and explain what would be happening in the next 20 minutes. They also had to get hold of the head of communications, Hege Tunstad. May-Britt texted her: "Can we have a chat?"

When Tunstad came into the office and saw their serious faces, she feared there had been an accident.

"They're going to get a Nobel Prize," Slørdahl whispered.

"Huh?" Tunstad said, as the conversation she'd had with Slørdahl earlier that morning raced through her mind. What about the unwritten rule that the prize never went to the same field of science two years in a row? She felt her heart pounding.[74] Would she be able to cope with the media storm? She rang NTNU's head of communications and said he had to drop everything and line up as many people as possible. Then they set up direct streaming from Stockholm on Slørdahl's PC.

The TV image showed an empty venue decorated for a solemn occasion. At eleven-thirty, the committee entered the hall, and its leader, Göran Hansson, positioned himself by the microphone. May-Britt watched the screen, head half-hidden in her hands as if to protect herself from what was to come. Hansson read from a sheet of paper in his hand: "The Nobel Assembly of the Karolinska Institute has today decided to award the 2014 Nobel Prize in Physiology or Medicine with one half to John O'Keefe and the other half jointly to May-Britt Moser and Edvard Moser for their discoveries of cells that constitute a positioning system in the brain."

May-Britt, who had just become one of Norway's two first Nobel prizewinners in Physiology or Medicine didn't say a thing. She was too overwhelmed to utter a word. Slørdahl gave her a hug. Since the news was now official, he threw open his office door, went out waving his arms above his head, and shouted: "They won a Nobel Prize!" People poured out of their offices to congratulate May-Britt, and the whole chaotic crowd made its way up the two levels to the fourth floor where the Moser group was based. The news had already reached the people up there, and the moment May-Britt walked in through the door, she was greeted with cheers and hugs. The Nobel prizewinner had got over her initial shock and was almost bubbling over with happiness. Hands above her head, she performed a victory dance along the narrow corridor.

Meanwhile, Edvard had yet to receive the message. He was somewhere over northern Germany with his cellphone switched off, still engrossed in his article.

At the same time, Professor Paola Pedarzani – Edvard and May-Britt's old friend from their student days in Oslo – was lecturing students in the old Anatomy Building on Gower Street, London. The day's topic was evaluating scientific articles, and Pedarzani had chosen the Mosers' 2005 paper about grid cells as the starting point for a critical review.[75] She and her students had picked apart the entire paper, point by point, to understand how the data supported the conclusions. After thoroughly examining each section, they concluded that it was a masterly piece of work.

As Pedarzani was rounding off her lecture, her cellphone rang. She saw it was her husband's number and picked up. She knew he would only ring if it were important. He told her that May-Britt and Edvard Moser had won the Nobel Prize along with her own colleague, John O'Keefe. She whooped for joy, then took the phone away from her ear and shouted out to the students that the scientists who wrote the paper they'd been reading had just won the Nobel Prize. "You see? I told you it was a good paper," she laughed.

At twelve-thirty, one hour after the rest of the world found out that he had won a Nobel Prize, Edvard's plane landed in Munich. Outside, a woman from the German aviation authority was waiting for him with flowers and a car to transport him through the airport area. She told him he'd won a prize, but when he asked which one, she wasn't sure. She thought maybe it was an award from the Max Planck Institute. Puzzled, Edvard switched on his cellphone to find out what had happened. At once, the messages started flooding in. During his flight alone, he had received 150 emails and 75 text messages. One of the first messages was from Göran Hansson. "Ring me at once. Important! Göran H."[76] Edvard's initial thought was the same as May-Britt's: Maybe he would be asked to comment on the fact that somebody he knew had won the Nobel Prize.

As Edvard was being transported by car to meet the delegation from the Max Planck Institute who were waiting for him at the airport, the circus was breaking loose at the Kavli Institute. Press and employees were invited to a celebration there at one o'clock. The Institute had been transformed into a self-running events bureau. Somebody had wheeled in a shopping trolley full of champagne: They had borrowed Per Andersen's tradition of celebrating milestones, although they'd diverged from his habit of offering port. And never before had the Institute had greater reason to celebrate. Some colleagues had bought cakes from a local bakery, and others had got hold of large quantities of sushi. The place was full of bouquets of flowers, and pretty soon it was packed with people.

A dozen journalists surrounded May-Britt with microphones, lights, and TV cameras. Champagne corks popped, and May-Britt talked non-stop into microphones and her telephone. As the cameras rolled, Norway's prime minister,

Erna Solberg, rang to congratulate her. When May-Britt finally got hold of Edvard, she was live on air with the national broadcaster, NRK. She had to concentrate to hear him over the hubbub and the surrounding journalists.

And so the day went on, amid radio, TV, and newspaper interviews with national and foreign journalists. The widely read German newspaper, *Frankfurter Allgemeine Zeitung*, had sent a team and requested a tour of the lab as well as 20 minutes alone with the Nobel prizewinner. May-Britt also had a call from NRK's daily news program, *Dagsrevyen*, inviting her and Edvard to make a guest appearance at their Oslo studio. She passed the cellphone to Slørdahl, who said it would be impossible for them to get to Oslo. Instead arrangements were made for her to be interviewed from a studio in Tyholt, NRK's Trondheim base, and Edvard from a studio in Munich.

The main story on *Dagsrevyen* that evening was about a Norwegian doctor who had been infected with Ebola fever and had been flown home to Norway for treatment. The rest of the program was dominated by the incredible scientific achievements in Trondheim.[77] But viewers could have been forgiven for thinking that the University of Oslo was responsible for securing the Nobel Prize for Norway: Even though NTNU chancellor and physician Gunnar Bovim was just a few hundred meters away from the *Dagsrevyen* studio in Oslo that day, they opted to invite Ole Ottersen, chancellor of the University of Oslo, onto the program instead. As a result, nobody got to hear NTNU's thoughts about how they had managed to win Norway's first Nobel Prize in Physiology or Medicine.

After an almost endless round of newspaper, radio, and TV interviews, May-Britt and her daughter Ailin rounded off the evening at Credo restaurant, where Chancellor Bovim and a team of close collaborators fêted the Nobel prizewinner. At the same time, on the eleven o'clock news, reporter Hallvard Sandberg, hinted that NTNU could lose its brand-new Nobel prizewinners to one of the leading American universities.[78]

THE NOBEL STRATEGY

The truth was that the Mosers had long been receiving plentiful offers from prestigious universities. But they had chosen to stay in Trondheim. How had NTNU managed to hold onto them for 20 years? And how was it even possible for a tiny university in a remote corner of the globe to win the world's most prestigious science prize?

It was incredible but not a matter of chance.

"I think the most important job I did as dean was keeping them in Trondheim," says Stig Slørdahl, who, as dean of the Faculty of Medicine, was their manager for ten years.[79] "Every single day, it was my job to fight for them to stay in Trondheim. We couldn't compete with the foreign universities on salary. But we could give them top research facilities – in that respect we could be just as good as the rest. And if you're treated well, you develop a loyalty toward your institution."

After the discovery of grid cells in 2005, the management of NTNU and the Faculty of Medicine realized that the Moser group was heading toward something at the Nobel level. Slørdahl didn't sit around waiting. NTNU and the Faculty of Medicine decided to offer the Kavli Institute special treatment, giving them what Slørdahl called a "gold card." This meant that the faculty ensured they got better service from the administration than other research groups. Hiring and finance issues would be dealt with swiftly and seamlessly. This was preferential treatment, and they made no secret of it. And the Moser Institute didn't just get extra service at the faculty level. When, at one point, they needed a new animal lab that would cost several tens of millions of kroner, former chancellor Torbjørn Digernes and the then head of finances made sure they got the money.

What's more, Dean Slørdahl deliberately started to put them forward for international prizes.

And the awards came flooding in. In 2008 they won the Fernström Prize for outstanding medical research. In 2011 they received the Louis-Jeantet Prize for Medicine and an Anders Jahre Prize for Medical Research. Two years later, they won the Louisa Gross Horwitz Prize and the Fridtjof Nansen Award for outstanding research. In 2014, they won the Karl Spencer Lashley Prize and the Körber Prize, and were elected foreign associates of the prestigious US National Academy of Sciences.

Slørdahl was especially pleased when the Mosers won the Louis-Jeantet Prize. He believes this was crucial to their winning the Nobel Prize later, as the chairman of the committee that awarded this prize was also secretary of the Nobel Committee for Physiology or Medicine.

A couple of days after the announcement of the Nobel Prize – and not without schadenfreude – Slørdahl had an opportunity to teach his institution's big brother in the south, the University of Oslo, what they could learn from the NTNU's success. In an editorial in Norway's business daily, *Dagens Næringsliv,* Slørdahl wrote that the NTNU had done precisely what the University of Oslo was advised to do in a report titled *Build a Ladder to the Stars*. In this report, renowned international academics had offered advice on how the university could make it to the top by 2020. "A recurring theme in the report is the importance of committing more wholeheartedly to individual scientists and research groups with high potential and giving them freedom and money. In other words, behave more like the Faculty of Medicine at NTNU: hand out gold cards to the best," wrote Slørdahl.[80]

The paradox was that the University of Oslo had "raised" NTNU's most successful researchers. What did NTNU think about that?

"They let them go. We couldn't afford to be that arrogant," Slørdahl says.

THE NOBEL CIRCUS CONTINUES

If anybody had been nervous about losing the Nobel prizewinning couple to a foreign country, the people of Trøndelag at least were reassured by the front page of Trondheim's local newspaper, *Addresseavisen*. "Promises to stay at NTNU," ran the headline above a photo of May-Britt Moser at the dinner with NTNU management the previous evening.

Edvard hadn't planned to come back from Germany so soon. "This will soon blow over," he'd told Slørdahl on the phone. But Slørdahl ordered him to take a flight back to Norway the next day: The Nobel circus showed no sign of abating – on the contrary, it was continuing at full tilt. It started early the next morning, when May-Britt Moser and Erna Solberg appeared on Norway's breakfast TV show *God morgen Norge*. The Foreign Minister, Børge Brende, made a trip to Trondheim to present a huge bouquet of flowers. Norwegian and foreign journalists from radio, TV, and the newspapers asked questions.

Later that day, Edvard returned to the city. For the second day in a row he was met off the plane with flowers and a huge hullabaloo. This time, the people waiting for him were journalists, TV crews, and a whole busload of colleagues waving Norwegian flags.

Once the initial pandemonium started to die down, requests came in for profile features and "at home" interviews. Most people seemed to be interested in the fact that they were a couple. When and how had they met? Edward and May-Britt dug out old photos of themselves as young students on country hikes and boat trips in South America, and of their engagement on top of Mount Kilimanjaro.

What very few people knew at this point was that despite their success as colleagues, the couple were no longer as close in their private life. How are you supposed to answer questions about your relationship as a couple when that relationship is coming to an end?

"We are fantastically good colleagues, we share something totally unique, we have shared values and complement each other very well – all this means that together we are dynamite. That was what we focused on when we got the Nobel Prize: that we were so proud – not just of what Edvard and I have achieved, but of what has, in a way, been built by us, around us," May-Britt says.

"We really thought the focus should be on the research we had done. But of course an awful lot of people were interested in the private aspect and wanted to write more about it. But I think that no matter what stage we were at in our lives, we never would have been open to that kind of thing. We have always turned down "at-home" interviews. It may be nice at the time, but later on you'll suddenly catch yourself in a contradiction," Edvard Moser says.

THE NOBEL DRESS AND THE KING OF SWEDEN

Two months later, on December 10, 2014, May-Britt was going on stage at the Stockholm Concert Hall clad in a long, dark blue dress and high heels. The dress was by Matthew Hubble, an English tunnel engineer turned designer. He had embroidered the dress with a beautiful silver pattern of neurons in honor of the Nobel prizewinners' research. The pattern and the designer's unique background generated many newspaper headlines and considerable attention in the weeks leading up to the ceremony. May-Britt was met on stage by King Carl Gustaf, who presented her with the Nobel medal and a Nobel plaque as trumpet fanfares sounded from the mezzanine. May-Britt shook the king's hand and thanked him. Then she bowed first to him, then to the Nobel committee, and, last of all, to the public. The TV cameras zoomed in on her daughters, Ailin and Isabel, who smiled and clapped, eyes fixed on their mother. After that, Edvard went up, dressed in tails, and received his medal to the same trumpet fanfares. "This is one place that will definitely be preserved by their spatial memory," said the commentator.[81]

When this highly memorable day and evening were over and it was time to leave Stockholm Concert House, May-Britt took King Carl Gustaf's place-card as a souvenir. All the other events and details they had experienced throughout the day were being transformed into dark silver nitrate crystals on the photographic film of their brains: the solemn entry, led by female students wearing tasseled student caps; the TV cameras; King Carl Gustaf; the Nobel medal; the trumpet fanfares; and the smiling faces of Isabel and Ailin. From now on, these memories would be linked to this place in the world and would, in turn, be inextricably linked to their identities.

CHAPTER 15

Into Whiteness – When Memories Die

Composer Bertil Palmar Johansen sits in an office at the Kavli Institute listening to a crackling from the loudspeakers. He has heard many neural recordings, and at first he doesn't notice anything special about the cell that PhD student Øyvind Høydal is playing back to him. The activity follows the same types of patterns he is used to hearing from these neurons.

Johansen first met May-Britt and Edvard when he was commissioned to write a piece of music to celebrate their Nobel Prize. Since then he and Moser had been collaborating to produce music based on the sounds from the laboratory.

But then something happens. The cell activity seems to be accelerating. "What's happening? – it's going bananas," Johansen thinks. It's as if it is boiling over. And then, abruptly, it stops. The cell falls silent. "This is a dying cell," Høydal tells him. But then the composer hears something else: a faint vibration of electrical signals. The sound reminds him of the static you hear between channels on an old-fashioned radio. It's as if the cell is being drained of electricity, a lonely sound that makes him think of outer space. And then, suddenly, there's a bang, as if somebody has fired a pistol shot. Johansen jumps. Then it goes quiet. Totally quiet.

The sense of infinite solitude and silence makes Johansen feel as if he is somewhere on the dark side of the moon. He sits there, hands in his lap, just listening for a while. Then he grasps it: This is what Alzheimer's is. This is how dramatic it is when the neurons that preserve our memories fall silent, leaving emptiness behind them.

And with that, the idea for a piece of music about Alzheimer's is born.

Many millions of people worldwide live with Alzheimer's disease. Nobody knows what causes it, and so there is no cure for it. What we do know is that neurons die and that cell death often starts in the entorhinal cortex and the hippocampus – precisely the areas that are important for our sense of place and memories. The brain of a person with Alzheimer's gradually forms huge empty spaces where once there were cells capable of encoding new memories. Slowly but surely, the patients lose their memories, and their families lose a loved one – for without our memories, we are no longer the person we once were. Scientists predict that ever-increasing numbers of us will eventually succumb to the disease as the average lifespan increases.

"Developing a brain disease is awful not only for the person who falls sick, but also for those around them. The fact that the child has to take charge of things for their own parent and say, you can't drive a car any more now – it's really hard, emotionally. And the disease demands huge resources from family, friends, and society," May-Britt says.

With all their knowledge of the brain, May-Britt and Edvard feel duty-bound to contribute to increasing knowledge about brain diseases, while continuing with their bench research. Their status as Nobel prizewinners makes it easier for them to take on high-risk projects. What's more, there turns out to be a direct link between grid cells and Alzheimer's.

When Edvard, May-Britt, and their colleagues in Trondheim discovered grid cells in rats, they assumed that humans had similar cells, even though this hadn't yet been proved. Five years later, researchers showed that humans too rely on grid cells to move around in the world. Neuroscientist Christian Doeller and his colleagues at University College London used a functional MRI scanner whose high resolution allowed it to read neural activity that correlated with the grid cells of human subjects as they moved around in a virtual city.[82] Doeller found that the people who were best at connecting events to places in the game were the ones whose grid cells produced the most regular hexagonal patterns. This indicated that grid cells play an important role in human memory. And that wasn't all.

Sense of place is one of the first faculties affected when people get Alzheimer's. Consequently, Doeller was interested in investigating whether people who were genetically predisposed to developing Alzheimer's had an imperfect grid pattern. In a new study, this time in collaboration with German and Dutch researchers, he compared the grid patterns in healthy people and people with a genetic mutation known as APOE-4, which increases the risk of developing Alzheimer's. These test subjects were also asked to move around in a virtual landscape, where they were tasked with putting objects in their correct places. The landscape looked like an open plain inside a volcanic crater. Although the test subjects were in their twenties, and none had yet developed Alzheimer's, the researchers could see a clear reduction in their grid patterns compared with the grid patterns of people without the genetic mutation.[83] The risk group also seemed to avoid moving out into the open plain and mostly kept to the edges, as though they had to rely on landmarks to find their way. In other words, those of us who produce less perfect grid patterns of the world may be at greater risk of developing Alzheimer's later in life.

"This may mean that you can go in and be diagnosed with Alzheimer's before developing symptoms. And if you want to combat the disease, you need to go in at an early enough stage that the cells aren't yet dead. If there's any way our center can help prevent people from developing this terrible disease, that would be a victory," May-Britt says.

In 2016 the Mosers recruited Christian Doeller to lead the project to translate their animal research into research on human beings. His recruitment was made possible in part by a large financial gift from Pauline Braathen, a billionaire widow who had lost her husband to Alzheimer's.

The first time May-Britt played the sound of a grid cell to Bertil Palmar Johansen, he was enthusiastic and asked to take the recordings home with him to do more work on them. It took him half a year to translate the sound recordings into rhythms and notes. But the first time he played them on the computer, with different instruments reproducing the rhythms of the different cells, he was surprised by how exciting it sounded. It felt as if time were standing still. At the same time, he was moved by the fact that something initially so remote from music – the cool electrical activity of cells deep within the brain – could sound so beautiful and rhythmical. Johansen sat in his office and laughed out loud. Then he developed it into a piece he called *My Running Rat*, which May-Britt took with her to Chicago and played to a lecture hall full of students, with great success.

The cell death recording Johansen had taken with him this time was dramatic and therefore easy to recreate as music. He immediately imagined violins playing the faint electrical vibration of the dying cell *glissando* – gliding movements where one note merges into the next. He also needed an audio image of a memory. Racking his brains for something appropriate, he suddenly remembered a recording he had nearly forgotten.

In the attic at his childhood home, Johansen had found an old recording of his grandmother playing the organ and singing the hymn *Long Have I Wandered.* It's a crackly sound recording from 1959, but the beautiful heartfelt voice of his grandmother shines through:

> Long have I wandered far from Thee,
> Father may I come home?
> Weary I am of this perilous road,
> I want to find my way.
> May I be thy servant?
> Take me by the hand!

The recording is a reminder of a woman who died many decades ago and who meant a great deal to the people who loved her. Johansen's grandmother lived on the island of Haramsøya, the island where Edvard's family spent their early days in Norway, and the organ she played may have been made by Edvard's father, who was an organ maker. Some years before, Johansen had digitized the recording and sent it to his siblings. He thought this was the kind of memory that would be strong enough to endure for a very long time, even when cell death has almost run wild. His own mother, aged 93, often speaks of her mother and remembers the hymns she played and sang. He knew that this memory would survive right until life's end. Even though it was painful to hear, he decided to cut the song to the rhythm of the dying cell and let the memory disintegrate into the music.

On June 22, 2017, May-Britt stands on the stage of Trondheim Spektrum with the musicians of the Trondheim Soloists. In the darkened concert hall sit more than 2,000 people, both locals and people who have traveled from all over the world to take part in the Starmus Science Festival, which brings together researchers, artists, and musicians. Accompanied by the musicians, May-Britt shows the audience how place cells react to a given location in the room, illustrated with the help of a violin in the orchestra, which plays every time May-Britt goes to a certain spot on the floor. She explains how place cells help us find our way, and how memories are linked to place. And then the time comes for the performance of Bertil Palmer Johansen's piece, *Into Whiteness*, in which the sound of his grandmother's voice from the almost 60-year-old recording is accompanied by the musicians on stage.

> May I be
> Thy servant
> Take me by
> The
> hand

Gradually, the song breaks down into tiny fragments that become fainter and fainter. Soon all that is audible is the *glissando* of the violins as they bow the electrical sound of a dying cell being sapped of its energy. In the end, the audience hears only the shriek of the violins, dwindling in number to two, then finally just one, thin gliding sound – until at last that lone violin, too, falls silent. And then comes a crash on the string of a double bass. It is over. The memory has been erased, along with one of the cells that preserved it. Everything is white.

Acknowledgements

Memory is a reconstruction, as I've been reminded by many of the scientists who contributed to this book. The book in your hands has been put together from the memories of several dozen people, and the whole thing is filtered through me. The result is, of course, not a full and complete account of what happened, but I have tried to weave everything together as truthfully as possible into a cohesive and exciting story. Memories have diverged more in some places than others, and in such cases, I have tried to draw on as many sources as possible to gain clarity about what actually happened.

This has been a demanding book to write. It was important for me to document the research processes as accurately and in as much detail as possible to give the reader a good image of how Nobel Prize-winning research is done. At the same time, I have tried to make scientific topics accessible to a general readership. This is easier said than done, given that my own field of expertise is chemistry, not neuroscience.

Many times along the way I felt like a rat in a Tolman maze, with a view of only one part of the maze at a time and incapable of seeing how the whole thing looked from above. I hit dead-ends, and now and then I sat down in a corner and despaired. But after exploring the maze for two years, I feel that I have now gained something of an overview at least. I hope I have managed to convey this insight to the reader.

I want to thank many people for making this book possible. First and foremost, I want to thank my editor, Anne Arnesen Mørch, who insisted that I write this book and wouldn't give up until I agreed to do it. She has since followed the entire project with the tenacity of a sheepdog, always helping me keep to the right narrative path. Thanks also to Håvard Parr, who stood in for Anne when she was on maternity leave. He ensured that the book project continued to progress and made many helpful contributions.

Thanks to the Norwegian Non-Fiction Writers and Translators Association, which gave me the funding that allowed me to spend six months writing full-time. That wasn't enough time to finish the book, but it was an important start. And thanks to Ruth Grüters, who was generous enough to give me time off from my job so that I could write, even though I had only just started work in the Department of Teacher Education. Without this time, I would not have been able to write the book.

Thanks to May-Britt and Edvard Moser for agreeing to collaborate with me and opening doors for me. You found time for interviews when I needed more information and impressed me with your rapid reviews and commentary on the manuscript.

Thanks to everybody who has given up their time to be interviewed, show me around laboratories and operation rooms, answer questions, and give me tips about topics and about people I ought to spend more time speaking with: Per Andersen, Per Brodal, Eva Aaboen Hanssen, Knut Rekdahl, Vidar Jensen, John O'Keefe, Carol Barnes, Jan Dyrstad, Jon Lamvik, Arne Valberg, Gunnar Bovim, Eivin Hiis Hauge, Marianne Fyhn, Sturla Molden, Bill Skaggs, Hill-Aina Steffenach, Vegard Brun, Stig Hollup, Kyrre Haugen, Ann-Mari Amundsgård, Trygve Solstad, Francesca Sargolini, Emilio Kropff, Hanne Stensola, Tor Stensola, Karel Ježek, Boleslaw Srebro, Siv Eggen, Eleanor Maguire, Bertil Palmar Johansen, Rita Elmkvist Nilsen, Paola Pedarzani, Randolf Menzel, Sturla Krekling, Morten Raastad, Stig Arild Slørdahl, Menno Witter, Terje Lømo, Øystein Orten, Pål Kvello and Annette Lykknes. Thanks to Ingvild Hammer for the loan of her photographs. An extra thank you to Richard Morris, who arranged a visit to Appleton Tower (even though it was closed for refurbishment) and gave me a tour of Edinburgh's old town. My apologies if I have left anybody out!

And my apologies to all the people I ought to have spoken to and whom I didn't have room to mention in the book. There just wasn't enough time!

Last but not least, my thanks to my family! You put up with me spending precious evenings and weekends of our brief time together here on Earth to sit poring over my PC instead of being with you. Thank you Tormod, Åsa, and Håvard – I love you all so much!

Notes

1. *Narrative of an expedition to the polar sea, in the years 1820, 1821, 1822, & 1823, Commanded by Lieutenant, Now Admiral, Ferdinand von Wrangel, of the Russian Imperial Navy*, Edited by Major Edward Sabine, R.A. F.R.S, London 1840, p. 146.
2. Darwin, 1873.
3. Quotes from Edvard Moser transcribed from interviews with Unni Eikeseth unless other sources are specified. Interview dates: April 1, 2014, May 27, 2014, March 18, 2016, April 8, 2016 and October 6, 2017.
4. *Scientific American*, 1979.
5. Kandel, 2006.
6. Eric Kandel was awarded the Nobel Prize for Physiology or Medicine in 2000.
7. Larry Weiskranz is the scientist responsible for this quote, which was printed in Alf Brodal's review article about the hippocampus. The information was drawn from Crompton, Alistair, "Hippocampus and the Sense of Smell. A Review, by Alf Brodal". Brain 2010: 133, 2509–2513.
8. Brown og Schäfer, 1888.
9. Victor, 1961.
10. Corkin, 2013, p. 19–33.
11. Squire, 2011.
12. NRK TV, 1980.
13. Andersen, 2006, p. 22.
14. Lømo, 2003.
15. Lømo, 2016.
16. Morris, Anderson, Baudry and Lynch, 1986.
17. Quotes from May-Britt Moser drawn from interviews with Unni Eikeseth unless other sources are named. Dates of the interviews: November 26, 2015, February 4, 2016, May 12, 2016, 9 August 2016, and September 26, 2017.
18. Per Andersen in interview with Unni Eikeseth, September 28, 2015.

19. Source: Kenneth Hugdahl, from his presentation of Edvard Moser at the conference Psykisk helse og rus in Oslo, 2 February 2015.
20. Moser, Mathiesen and Andersen, 1992, p. 1324.
21. Moser, Trommald and Andersen, 1994, p. 12673.
22. Røed and Snare, 1995, p. 18.
23. Bear, Connors and Paradiso, 2007.
24. Morris, Anderson, Baudry and Lynch, 1986.
25. John O'Keefe in interview with Unni Eikeseth, September 11, 2014.
26. Nobel Prize.org, 2014a.
27. Sweet, 2014.
28. Nobelprize.org, 2014a.
29. Nobelprize.org, 2014b, p. 276.
30. Tolman, 1948.
31. O'Keefe and Dostrovsky, 1971.
32. O'Keefe and Nadel 1978.
33. Carol Barnes in interview with Unni Eikeseth, April 14, 2016.
34. Boleslaw Srebro in interview with Unni Eikeseth, February 17, 2016.
35. Moser, Krobert, Moser and Morris, 1998.
36. Posluszna, 2015, p. 77.
37. Jan Dyrstad in interview with Unni Eikeseth, September 14, 2015.
38. Vegard Brun in Skype interview with Unni Eikeseth, April 15, 2016.
39. Brun, Otnæss, Molden, Steffenach, Witter, Moser and Moser, 2002.
40. Ramón y Cajal, 1902.
41. Quirk, Mueller, Kubie and Ranck, 1992.
42. Frank, Brown and Wilson, 2000.
43. Menno Witter in interview with Unni Eikeseth, September 30, 2015.
44. Fyhn, Molden, Witter, Moser and Moser, 2004.
45. Redish and Touretzyk, 1996.
46. Bill Skaggs in an e-mail to Unni Eikeseth, June 3, 2017.
47. Marianne Fyhn in interview with Unni Eikeseth, May 6, 2014.
48. Reproduction of the contents of the e-mail Bill Skaggs sent to Edvard Moser in October 2004, as communicated in an e-mail from Bill Skaggs to Unni Eikeseth, October 19, 2017.
49. Samsonovich and McNaughton, 1997.
50. Wikipedia, 2017.
51. Hafting, Fyhn, Molden, Moser and Moser, 2005.
52. NRK TV, 2015.

53. Sargolini, Fyhn, Hafting, McNaughton, Witter, Moser and Moser, 2006.
54. Francesca Sargolini in Skype interview with Unni Eikeseth, 3 June 2016.
55. Hartley, Burgess, Lever, Cacucci and O'Keefe, 2008.
56. Trygve Solstad in interview with Unni Eikeseth, 24 May 2016.
57. Solstad Bocarra, Kropff, Moser and Moser, 2008.
58. Emilio Kropff in Skype interview with Unni Eikeseth, 24 May 2016.
59. Kropff, Carmichael, Moser and Moser, 2015.
60. Yates, 1966/1999.
61. O'Keefe and Speakman, 1987.
62. Kubie, 2013.
63. Leutgeb, Leutgeb, Barnes, Moser, McNaughton and Moser, 2005.
64. Moser, Rowland and Moser, 2015.
65. Karel Ježek in Skype interview with Unni Eikeseth, 23 October 2017.
66. Ježek, Henriksen, Treves, Moser and Moser, 2011.
67. Hanne Stensola and Tor Stensola in interview with Unni Eikeseth, June 17, 2016.
68. Stensola, Stensola, Solstad, Frøland, Moser and Moser, 2012.
69. Tulving, 2002.
70. Zinkivskay, Nazr & Smulders, 2009.
71. Allen and Fortin, 2013.
72. Nobelprize.org, 2014d.
73. Slørdahl, October 10, 2014.
74. Tunstad.
75. Paola Pedarzani in interview with Unni Eikeseth, September 9, 2015.
76. Nobelprize.org 2014c.
77. NRK TV, October 6, 2014.
78. NRK TV, October 6, 2014b.
79. Stig Slørdahl in interview with Unni Eikeseth, June 28, 2016.
80. Slørdahl, October 17, 2014.
81. NRK TV, December 10, 2014.
82. Doeller, Barry and Burgess, 2010.
83. Kunz, Schröder, Lee, Montag, Lachmann, Sariyska, Reuter, Stirnberg, Stöcker, Messing-Floeter, Fell, Doeller and Axmacher, 2015.

Bibliography

Andersen, P. (2006). Inhibitory Circuits in the Thalamus and Hippocampus – An Appraisal after 40 Years. *Progress in Neurobiology*, 78.

Allen, T.A. & Fortin, N.J. (2013). The Evolution of Episodic Memory. *Proceedings of the National Academy of Sciences*, 110, 10379–10386.

Bear, M.F., Connors, B.W. & Paradiso, M.A. (2007). *Neuroscience. Exploring the Brain*. Lippincott Williams & Wilkins.

Brown, S. & Schäfer, E.A. (1888). An Investigation into the Functions of the Occipital and Temporal Lobes of the Monkey's Brain. *Philosophical Transactions of the Royal Society of London. B*, 9.

Brun, V.H., Otnæss, M.K., Molden, S., Steffenach, H.-B., Witter, M.P., Moser, M.-B. & Moser, E.I. (2002). Place Cells and Place Recognition Maintained by Direct Entorhinal-Hippocampal Circuitry. *Science*, 296 (5576), 2243–2246.

Corkin, S. (2013). *Permanent Present Tense*. New York: Basic books.

Crompton, A. (2010). The Hippocampus and the Sense of Smell. A Review, by Alf Brodal. *Brain*, 1947: 70, 179–222. *Brain*, 133:9, 2509–2513.

Darwin, C. (1873). Origin of Certain Instincts. *Nature*, 7, 417–418. Taken from: http://darwin-online.org.uk

Doeller, C.F., Barry, C. & Burgess, N. (2010). Evidence for Grid Cells in a Human Memory Network. *Nature*, 463, 657–661.

Frank, L.M., Brown, E.N. & Wilson, M.A. (2000). Trajectory Encoding in the Hippocampus and Entorhinal Cortex. *Neuron*, 27, 169–178.

Fyhn, M., Molden, S., Witter, M.P., Moser, E.I. & Moser, M.-B. (2004). Spatial Representation in the Entorhinal Cortex. *Science*, 305 (5688), 1258–1264.

Hafting, T., Fyhn, M., Molden, S., Moser, M.-B. & Moser, E.I. (2005). Microstructure of a Spatial Map in the Entorhinal Cortex. *Nature*, 436, 801–806.

Hartley, T., Burgess, N., Lever, C., Cacucci, F. & O'Keefe, J. (2000). Modelling Place Fields in Terms of the Cortical Inputs to the Hippocampus. *Hippocampus*, 10, 369–379.

Jezek, K., Henriksen, E., Treves, A., Moser, E.I. & Moser, M.-B. (2011). Theta-Paced Flickering between Place-Cell Maps in the Hippocampus. *Nature*, 478 (7368), 246–249.

Kandel, E.R. (2006). *In Search of Memory: The Emergence of a New Science of Mind.* New York: W.W. Norton & Company.

Kropff, E., Carmichael, J., Moser, M-B. & Moser, E.I. (2015). Speed Cells in the Medial Entorhinal Cortex. *Nature*, 523, 419–424.

Kubie, J. (2013). *Place Cells, Remapping and Memory*. Taken [June 19, 2016] from: http://blog.brainfacts.org/2013/10/place-cellsremapping-and-memory/#.V2ZtiriLSUk

Kunz, L., Schröder, T.N., Lee, H., Montag, C., Lachmann, B., Sariyska, R., Reuter, M., Stirnberg, R., Stöcker, T., Messing- Floeter, P.C., Fell, J., Doeller, C.F. & Axmacher, N. (2015). Reduced Grid-cell–like Representations in Adults at Genetic Risk for Alzheimer's Disease. *Nature*, 350, 430–433.

Leutgeb, S., Leutgeb, J.K., Barnes, C.A., Moser, E.I., McNaughton, B.L. & Moser, M.-B. (2005). Independent Codes for Spatial and Episodic Memory in Hippocampal Neuronal Ensembles. *Science*, 309 (5734), 619–623.

Lømo, T. (2003). The Discovery of Long-Term Potentiation. *Phil. Trans. R. Soc. Lond. B*, 358, 617–620.

Lømo, T. (2016). Scientific Discoveries: What is Required for Lasting Impact. Annual *Review of Physiology*, 78, 2.1–2.21.

Moser, E.I., Mathiesen, I. & Andersen, P. (1993). Association Between Brain Temperature and Dentate Field Potentials in Exploring and Swimming Rats. *Science*, 259 (5099), 1324–1326.

Moser, E.I., Krobert, K.A., Moser, M.-B. & Morris, R.G. (1998). Impaired Spatial Learning after Saturation of Long-Term Potentiation. *Science*, 281 (5385), 2038–2042.

Moser, M.-B., Trommald, M. & Andersen, P. (1994). An Increase in Dendritic Spine Density on Hippocampal CA1 Pyramidal Cells Following Spatial Learning in Adult Rats Suggests the Formation of New Synapses. *Proceedings of the National Academy of Sciences of the United States of America*, 91 (26), 12673–12675.

Moser, M.-B., Rowland, D.C. & Moser, E.I. (2015). Place Cells, Grid Cells and Memory. *Cold Spring Harbor Perspectives in Biology*, 7:a021808.

Morris, R.G.M., Anderson, E., Baudry, M. & Lynch, G.S. (1986). Selective Impairment of Learning and Blockade of Long-Term Potentiation In Vivo by AP5, an NMDA Antagonist. *Nature*, 319, 774–776.

Nobelprize.org (2014a). *John O'Keefe – Biographical.* Taken [February 25, 2017] from:http://www.nobelprize.org/nobel_prizes/medicine/laureates/2014/okeefe-bio.html

Nobelprize.org (2014b). *John O'Keefe – Nobel Lecture: Spatial Cells in the Hippocampal Formation.* Taken [February 25, 2017] from http:// www.nobelprize.org/nobel_prizes/medicine/laureates/2014/ okeefe-lecture.html

Nobelprize.org (2014c). *Edvard Moser – Biographical.* Taken [February 26, 2017] from: http://www.nobelprize.org/nobel_prizes/medicine/laureates/2014/edvard-moser-bio.html

Nobelprize.org (2014d). *Interview with May-Britt Moser.* Taken [July 1, 2016] from http://www.nobelprize.org/nobel_prizes/medicine/laureates/2014/may-britt-moser-telephone.html

NRK TV (1980). *Husker du?* [Episode from TV series] Din fantastiske hjerne. Taken from: https://tv.nrk.no/serie/din-fantastiskehjerne/ FOLA00005380/13-05-1980

NRK TV (August 1, 2005). *Dagsrevyen.*

NRK TV (October 6, 2014a). *Dagsrevyen.* Taken from: https://tv.nrk.no/serie/dagsrevyen/NNFA19100614/06-10-2014

NRK TV (October 6, 2014b). *Kveldsnytt.* Taken from: https://tv.nrk.no/ serie/kveldsnytt/NNFA23100614/06-10-2014#t=6m15s

NRK TV (December 10, 2014). *Nobel Prize Award Ceremony.* Taken from: https://tv.nrk.no/serie/nyheter/NNFA41020414/10-12-2014

O'Keefe, J. & Dostrovsky, J. (1971). The Hippocampus as a Spatial Map. Preliminary Evidence from Unit Activity in the Freely-Moving Rat. *Brain Research*, 34, 171–175.

O'Keefe, J. & Nadel, L. (1978). *The Hippocampus as a Cognitive Map.* Oxford: Clarendon Press. Available online: http://www.cognitivemap.net/

O'Keefe, J. & Speakman, A. (1987). Single Unit Activity in the Rat Hippocampus during a Spatial Memory Task. *Experimental Brain Research*, 68, 1–27.

Posluszna, E. (2015). *Environmental and Animal Rights Extremism, Terrorism and National Security.* Oxford: Butterworth-Heinemann.

Quirk, G.J., Mueller, R.U., Kubie, J.L. & Ranck, J.B. Jr. (1992). The Positional Firing Properties of Medial Entorhinal Neurons: Description and Comparison with Hippocampal Place Cells. *Journal of Neuroscience*, 12 (5), 1945–63.

Ramón y Cajal, S. (1902). Sobre un Ganglio Especial de la Corteza Esfeno-Occipital. *Trab. Lab. Invest. Biol. Universidad de Madrid*, 1, 189–206.

Redish, A.D. & Touretzky, D.S. (1996). Cognitive Maps Beyond the Hippocampus. *Hippocampus*, 7, 15–35.

Røed, L.L. & Snare, K. (December 10, 1995). Bruk hodet! *Aftenposten*, 18–19.

Samsonovich, A. & McNaughton, B.L. (1997). Path Integration and Cognitive Mapping in a Continuous Attractor Neural Network Model. *Journal of Neuroscience*, 17, 5900–5920.

Sargolini, F., Fyhn, M., Hafting, T., McNaughton, B.L., Witter, M., Moser, M.-B. & Moser, E.I. (2006). Conjunctive Representation of Position, Direction and Velocity in Entorhinal Cortex. *Science*, 312 (5774), 758–762.

Scientific American (1979). *The Brain*. San Francisco: W.H. Freeman and Company.

Slørdahl, S. (October 10, 2014). *Nobeljubel.* Taken [July 1, 2016] from http://forskning.no/blogg/stig-slordahls-blogg/nobeljubel

Slørdahl, S. (October 17, 2014). *Gullkortteorien.* Taken from http://forskning.no/blogg/stig-slordahls-blogg/gullkortteorien

Solstad, T., Boccara, C.N., Kropff, E., Moser, M-B. & Moser, E.I. (2008). Representation of Geometric Borders in the Entorhinal Cortex. *Science*, 322 (5909), 1865–1868.

Squire, L. (2011). Annual Review of Physiology, 34, 259–288.

Stensola, H., Stensola, T., Solstad, T., Frøland, K., Moser, M.-B. & Moser, E.I. (2012). The Entorhinal Grid is Discretized. Nature, 492, 72–78.

Sweet, D. (October 16, 2014). *Nobel Winner Has Very Fond Memories of McGill.* Taken [February 25, 2017] from http://publications.mcgill.ca/reporter/2014/10/nobel-winner-has-very-fond-memoriesof-mcgill/

Tolman, E. (1948). Cognitive Maps in Rats and Men. *The Psychological Review*, 55 (4), 189–208. Taken [February 24, 2017] from: http://psychclassics.yorku.ca/Tolman/Maps/maps.htm

Tulving, E. (2002). Episodic Memory: From Mind to Brain. *Annual Review of Psychology*, 53, 1–25.

Tunstad, H. (October 9, 2014). *Noble ettertanker – eller den mest hektiske dagen i mitt liv.* Taken [July 1, 2016] from: https://hegetunstad.wordpress.com/2014/10/09/noble-ettertanker-eller-denmest-hektiske-dagen-i-mitt-liv/

Victor, M., Angevine, J.B., Mancall, E.L. & Fisher, C.M. (1961). Memory Loss with Lesions of Hippocampal Formation. *Archives of Neurology*, 5, 244–263.

Von Wrangel, F. & Sabine, E. (red.) (1844). *Narrative of an Expedition to the Polar Sea, in the Years 1820, 1821, 1822 & 1823.* London: James Madden & Co.

Wikipedia (2017). *Autocorrelation.* Taken from: https://en.wikipedia. org/wiki/Autocorrelation

Yates, F. (1966/1999). *The Art of Memory*. London: Routledge & Kegan Paul.

Zinkivskay, A., Nazir, F. & Smulders, T.V. (2009). What-Where-When Memory in Magpies (Pica pica). *Animal Cognition*, 12, 119–125.

Index